Contents

D0132590

Preface

The marriage of computer technology and traditional machining disciplines has given birth to revolutionary new machine tools. CNC turning and machining centres are now capable of doing the work of three or four conventional machine tools at a single setting. Exciting new concepts in the organisation of production are creating production facilities that can adapt to batch or one-off manufacturing operations with comparative ease and the minimum of human intervention.

The inevitable demand for skilled personnel to respond to the challenges of this computerised manufacturing technology is now gathering momentum. Dedicated college courses are being established, and company training policies are being re-formulated to reflect the importance now attached to these emerging technologies.

This book has been written to offer a comprehensive introduction to Computer Numerical Control to those readers wishing to learn the fundamentals of CNC as applied to metal cutting machine tools. It is envisaged that such readers will be:

- Students of engineering being introduced to CNC for the first time through vocational or short college-based courses.
- Established engineers wishing to update their knowledge of CNC technology.
- Busy engineering managers who wish to gain an insight into how CNC operates in the manufacturing environment.

Since CNC is not an end in itself, I have attempted, throughout the book, to discuss the subject within a context of manufacturing that emerges from the use of conventional machine tools through stand-alone CNC machine tools to the modern concepts of an integrated, unmanned manufacturing environment.

The text covers fully the syllabus content of established courses in CNC, at both advanced craft and introductory technician levels. In support of these study areas, related questions appear at the end of each chapter and a number of programming examples are included.

In writing this book I am indebted to the various people and organisations that have responded so willingly to my requests for photographs to appear within the text.

The book was produced using a BBC microcomputer and the View word processor. I should like to express my sincere thanks to Epson UK Ltd. for supporting the project by so generously providing an Epson RX-100 printer on which the manuscript was prepared. All the part programming segments within the text are reproduced directly from the output of this printer.

Last, but by no means least, I must thank my wife Helen and my daughter Laura for their infinite patience and caring support during the preparation of the book.

S. B. Leatham-Jones
Hoghton 1985

"For Laura and Stacey – with love, pride and happiness"

Acknowledgements

The author and publisher would like to extend sincere thanks to the following organisations who have contributed to the production of this book:

Aids Data Systems Ltd.
Cincinnati Milacron Ltd.
City and Guilds of London Institute
Dainichi Sykes Robotics Ltd.
Engineering and Scientific Equipment Ltd.
Epson U.K. Ltd.
PGM Ballscrews Ltd.
Sandvik U.K. Ltd.

Economic considerations 1

1.1 NC and CNC explained

1.1/0 What is numerical control?

Numerical Control (NC) is the technique of giving instructions to a machine in the form of a code which consists of numbers, letters of the alphabet, punctuation marks and certain other symbols. The machine responds to this coded information in a precise and ordered manner to carry out various machining functions. These functions may range from the positioning of the machine spindle relative to the workpiece (the most important function), to controlling the speed and direction of spindle rotation, tool selection, on/off control of coolant flow, and so on.

Instructions are supplied to the machine as **blocks** of information. A block of information is a group of commands sufficient to enable the machine to carry out one individual machining operation. For example, a block of information may command the machine to move the worktable to a specific coordinate position under rapid traverse, or set speed and feed values to carry out the machining of contours. Each block is given a **sequence number** for identification. The blocks are then executed in strict numerical order.

A set of instructions forms an **NC program**. When the instructions are organised in a logical manner they direct the machine tool to carry out a specific task—usually the complete machining of a workpiece or "part". It is thus termed a **part program**. Such a part program may be utilised, at a later date, to produce identical results over and over again.

1.1/1 NC machine tools

Automatic control of NC machine tools relies on the presence of the part program in a form that is external to the machine itself. The NC machine does not possess any "memory" of its own and as such is only capable of executing a single block of information, fed to it, at a time. For this reason part programs are normally produced, and stored, on punched tape.

To machine a part automatically, the **machine control unit** (MCU) will read a block of information, then execute that block, read the next block of information and execute that block, and so on. With the punched tape

1

installed it is also possible to "single step" a part program by instructing the machine tool to pause after the execution of each block. This is semi-automatic operation. It is also possible to enter data manually, by setting dials and switches, at the machine console. After each setting the machine will carry out the instruction and wait for the next setting. Thus, it is also possible to effect control by **manual data input** (MDI).

Since a part cannot be produced automatically without the tape being run through the machine block by block, they are often referred to as *tape-controlled machines*. The production of repetitive, identical parts thus relies on

a) the tape being present
b) the tape being in good condition.

Any number of identical parts being produced thus imposes harsh operating conditions on the punched tape and subsequent wear, especially of the feed holes used to transport the tape through the tape reader, is likely.

The features of early NC control systems, including any required options, had to be specified before purchase since, in most cases, they had to be built in as part of the hardware. They were known as *hardwired controllers*. It was difficult to add extra features at a later date and this made NC installations bulky and expensive. Since updating or upgrading was difficult, many NC machines soon became outdated and obsolete.

1.1/2   What is computer numerical control?

Computer Numerical Control (CNC) retains the fundamental concepts of NC but utilises a dedicated **stored-program computer** within the machine control unit. CNC is largely the result of technological progress in microelectronics (the miniaturisation of electronic components and circuitry), rather than any radical departure in the concept of NC.

CNC attempts to accomplish as many of the MCU functions as possible within the computer **software** which is programmed into the computerised control unit. This greatly simplifies the CNC hardware, significantly lowers purchase costs, and improves reliability and maintainability.

Updates and upgrades are relatively simple. In many cases it is only the stored **operating program** that needs to be modified. The main operating program is stored within the CNC control unit on a special memory chip. Any updates in the control system can be accomplished by replacing the chip with one containing the updated software. The memory chip removed can then be re-programmed with the current operating program. Any *circuit* modifications can be carried out with ease by simply replacing, or adding, components housed on a **printed circuit board** (PCB). Indeed, many modern electronic systems (from simple TV sets to sophisticated computer systems) are increasingly being constructed on a plug-in basis of electronic cards.

Modern CNC machines are thus tools with both current and future value. Obsolescence is, as far as possible, designed out.

CNC control units, like the computers on which they are based, operate according to a *stored program* held in *computer memory*. This means that part programs are now able to become totally resident within the memory of the control unit, prior to their execution. No longer do the machines have to operate on the "read-block/execute-block" principle. This eliminates the dependency on slow, and often unreliable, tapes and tape reading devices— probably the weakest link in the chain. Programs can, of course, still be loaded into the CNC machine via punched tape, but only one pass is necessary to read the complete part program into the memory of the control unit.

1.1/3   CNC machine tools

Many CNC machine tools still retain many of the constructional and physical design aspects of their NC counterparts. However, many new control features are made available on CNC machines, which were impossible, uneconomical or impractical to implement on early NC machines. Such new features include:

a) **Stored Programs**   Part programs may be stored in the memory of the machine. The CNC can then operate directly from this memory, over and over again. Use of the tape reader (and its unreliability) is virtually eliminated. For long production runs the part program may be retained in memory, even when the power is removed (say at the end of a shift or at a weekend), by the use of *battery back-up* facilities that keep only the memory supplied with power. Often, more than one program may be resident in the control unit memory at one time with the ability to switch between them.

b) **Editing Facilities**   Editing can be carried out on the part program held in memory. Thus, errors, updates, and improvements can be attended to at the machine. Such edits are stored in computer memory and override the tape information as read in. A new, and corrected, tape may then be punched directly from the CNC control unit. This ensures that the most up-to-date version of the part program is retained as current.

c) **Stored Patterns**   Common routines such as holes on a pitch circle, pocketing sequences, drilling and tapping cycles can be built in and retrieved many times. There is facility for user-defined sequences (such as roughing cycles, start-up routines, etc.) to be stored and retrieved in the same way. Only certain parameters have to be specified and the computer control will carry out the necessary calculations and subsequent actions.

d) **Sub-programs**   For repetitive machining sequences, sub-programs may be defined once and then be repeatedly called and executed as required. This considerably shortens part programs by eliminating the need to repeat sections of identical program code. For example, it may be required to machine the same set of holes but at a different position within the workpiece.

e) **Enhanced Cutter Compensation**   When a part program is written, it is normally done with a particular type and size of cutter in mind. The positioning of the cutter relative to the workpiece will need to take account of the dimensions of the cutter. It may be the case that, when the part program comes to be run on the machine, the particular cutter specified is not available. CNC control units allow "compensations" and "offsets" to be made for the differences in dimensions between the actual cutter and the specified cutter. Thus, the part program is now independent of the cutter specified when writing the program. This facility can also be brought into play in the case of tool breakage during the machining cycle, where different cutters may have to be reloaded to continue the machining sequence.

f) **Optimised Machining Conditions** The extremely fast response of computer technology, coupled with sophisticated calculation ability, enables machining conditions to be constantly monitored by the control unit. Spindle speed on a CNC lathe, for example, can be perfectly matched (and adjusted automatically) as the depth of cut varies. It is common to witness the spindle speed increase when a facing cut is taken from the outside diameter of a bar to its centre. Feed rate can be optimised by monitoring power consumed.

g) **Communications Facilities** The utilisation of computer technology within the CNC control unit offers the advantage of being able to communicate with other computer-based systems. Part programs may thus be downloaded from other host computers. Such host computers may be simple databases of different part programs, or sophisticated computer aided design systems.

h) **Program Proving Facilities** Many modern control systems contain software that will process the resident part program information and indicate the component shape that will be produced before machining takes place. This is often displayed graphically on a visual display unit (VDU) on the operating console.

i) **Diagnostics** Most modern CNC machines come equipped with comprehensive diagnostic software for the self-checking of its electronic operation. For example, there might be a diagnostic routine to check the operation of the memory chips. It would write a known test pattern into memory and then read it out again, checking it for validity. Any discrepancy could indicate a memory fault.

j) **Management Information** Since the CNC system controls nearly all functions from the resident computer, much useful information on machine utilisation can be accessed, i.e., spindle-on time, part run time, downtime, etc., can be logged and output to other computer systems or peripheral devices for subsequent reading and analysis.

In addition, many modern CNC machine tools are now capable of automatic tool changing without manual intervention. A number of standard (and/or specialised) cutting tools may be loaded into a rotating turret or carousel, and called up under the control of the part program.

1.2 Industrial applications of CNC

1.2/0 Machining

Undoubtedly the biggest application area for CNC control, at present, is in the field of machining. Indeed, it was the response to a machining problem that originally gave birth to the first CNC machine.

(A three-motion NC milling machine was successfully demonstrated at the Massachusetts Institute of Technology in 1952. It was developed as a result of problems encountered with the complex machining of curved aircraft components, to close accuracies, on a repeatable basis.)

So great is the influence of CNC in this area that revolutionary new machine tools are being developed to harness its potential. Machines such as Turning Centres or Machining Centres, which can accomplish a wide variety of machining operations at a single setting, are now commonplace. Furthermore, the development of these machine tools is realising exciting new concepts in the way in which production itself is being organised. Machining Cells, Flexible Manufacturing Systems, Integrated Manufacturing are all current developments that have been spawned by the wide influence of CNC.

A typical CNC machining centre is shown in Fig. 1/1 and a CNC turning centre is shown in Fig. 1/2.

CNC control is also being applied to the more specialised techniques of metal removal, such as grinding and electro-discharge machining (EDM).

1.2/1   Fabrication and welding

Close behind machining activities are applications in fabrication and welding. Since CNC is basically a machine control, and not a machining control, it matters little what the machine is.

For example, by substituting an oxy-acetylene, plasma, or laser cutting head for the machine tool spindle, the result can be an extremely versatile and productive means of *cutting* plate material. Replacing the cutting head with a *welding* torch enables CNC fabrication to be achieved. It must be conceded, however, that robotic welding techniques probably represent an even more versatile option in pure welding applications.

Folding and *shearing* machinery represent other application areas for CNC control in the fabrication field. CNC *bending* equipment in pipe and tube applications is making a significant impact in areas such as car exhaust pipe manufacture. A variety of complex bending patterns can be reproduced quickly and accurately making optimum use of material.

1.2/2   Presswork

In parallel with, and in support of, developments in CNC fabrication and welding applications, piercing, notching and nibbling applications have now developed under CNC control.

Blanking and *piercing* are operations carried out on sheet material whereby a suitably shaped punch is pressed through the material under heavy, and often impact, loads. In piercing, the piece of material punched out is scrap, and it is the component that is left.

Nibbling and *notching* consist of a reciprocating punch that repeatedly "nibbles" away at the material being fed underneath it. These processes are utilised where holes, or edge contours of a complex shape, are required within sheet material and the production of a suitably shaped punch would be technically or economically impractical.

Punching apertures and hole patterns is an ideal application for CNC control. It basically only requires precise positioning in two (X and Y) axes. The ability to punch shapes ranging from the most simple to the very complex, utilising just simple standard punches, is very attractive from a production engineering standpoint. Press speeds in excess of 100 hits per minute are common. Where a range of punch options is required, automatic punch changing can be provided. Tool turrets comprising 36 punch stations are common. Integral slug conveyors, running continuously, provide an efficient means of removing the punched slugs without interfering with the press operation.

In many cases the computer can also be utilised to produce optimum "nesting" patterns for components being blanked from sheet material. *Nesting*

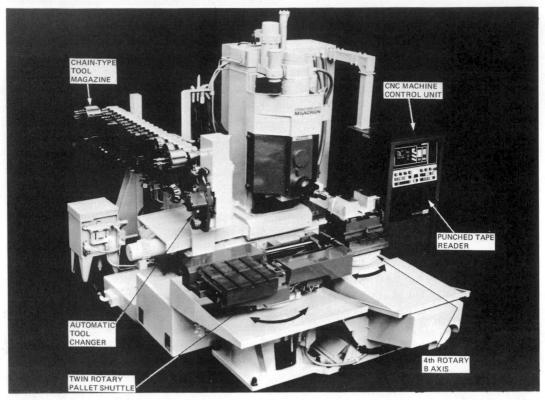

Fig. 1/1 A Machining Centre [*Cincinnati Milacron*]

Fig. 1/2 A Turning Centre [*Cincinnati Milacron*]

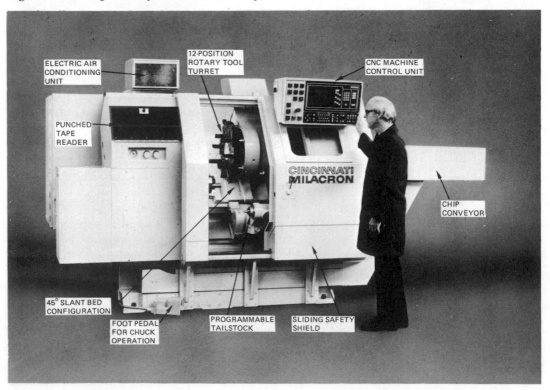

is the term given to the layout of shaped components, within the sheet material area, to give the maximum number of components per sheet.

CNC presswork eliminates the time-consuming operations of sheet layout and drawing interpretation. The operator is thus freed to carry out other functions. Material handling is minimised since, once loaded onto the machine, all component movement (and in many cases removal and stacking) is carried out automatically.

1.2/3 Inspection and measurement

Machine tools can be made to produce complex components under CNC control. The next logical step is the utilisation of a machine that can check and inspect the same components under CNC control. The basis of dimensional measurement or inspection has already been provided in the part program used to manufacture the components.

Three-dimensional coordinate measuring machines (CMMs) are the result. Such machines are not restricted to measuring individual components. Assemblies and sub-assemblies may be measured by probing and thus inspected by comparing the actual dimensions with the required dimensions. The ability to "remember" positions and dimensions, together with the ability to "re-state" datums, make these machines extremely versatile. Control software within the machines makes comprehensive information available to the operator via visual display units and printed hardcopy. Desired dimensions, actual dimensions, actual errors and their locations can be provided. Computational facilities can also take into account, and compensate for, any specified tolerances relating to component features. In specific circumstances, results from inspecting a first-off component may be fed back automatically to the machining process, allowing automatic adjustments of cutter path movements.

Most CMMs operate in a stand-alone mode since various problems exist in trying to incorporate them into a flexible machining system. Such problems include:

a) The machines are susceptible to vibration.
b) Strict environmental control has to be observed to minimise the effects of heat and humidity on dimensional and volumetric changes.
c) Components have to be allowed to reach thermal stability after the heating effects of the machining process.
d) Components have to be thoroughly cleaned before passing to the inspection stage. The effects of dirt or swarf can distort the results obtained from the inspection process.

Coordinate measuring machines can also be used for other tasks within the manufacturing system. For example, they can be used for scribing and marking-out operations. Rough castings or forgings can be probed to ensure that the unmachined as-cast or as-forged dimensions provide enough material for subsequent machining. They can also probe for the presence or absence of features such as cored holes. Sheet metal and pattern development layouts can also be accomplished more quickly and accurately on these machines than by other methods.

1.2/4 Assembly

Where components are required to be assembled within a flat plane using X and Y coordinates (obviating complex manipulative problems), CNC assembly machines may be viable alternatives to the use of robot devices.

An obvious example is in the electronics industry where discrete components and integrated circuits have to be assembled on printed circuit boards. Speed of operation and ease of programming become more important than the greater flexibility (and attendant clumsy operation) offered by robots, since thousands or millions of components may be involved. Many writers have observed the eerie phenomenon of computers building computers.

A second example concerns the assembly of wiring harnesses for motor vehicles. Different lengths of different coloured wire are threaded along precisely defined routes between previously set guiding pins. Manual operation of setting the different pin patterns for the different harness designs is a slow and tedious operation. Both pin setting and threading can be accomplished under CNC control.

1.2/5 Materials handling

CNC as applied to materials handling is widely exhibited in the form of the now familiar industrial robot. Whilst robots may not immediately be thought of as applications of CNC, much of the programming and control philosophies are directly related. Robots are becoming an important discipline in CNC-related applications since they offer the possibility of unmanned operation of manufacturing installations. Industrial robots are discussed in Chapter 8.

This book will concentrate on CNC technology as applied to metal cutting machine tools.

1.3 Economic benefits of CNC

1.3/0 CNC versus conventional machining

Conventional machining relies on a skilled operator to manipulate the machine tool handwheels to produce a required component. The operator has to examine the drawing many times, during the operation, to determine the dimensions that apply; and must decide (manually calculate) by how much each handwheel must be turned to produce the desired result. Since only one handwheel can be controlled at a time, with any degree of accuracy, contouring is limited unless special attachments are utilised. These attachments must be acquired, attached to the machine and duly set up. This can be a time-consuming activity in itself.

Very often, metal removal requires a series of cuts before the final result is achieved. Measurement of the part must be carried out in between these cuts. It is almost impossible to predict the final condition of the part, during multiple cuts, even though the handwheels have calibrated scales.

Because of limitations involved in the design of conventional machine tools, much tool changing, tool setting and workpiece re-setting is often involved during the machining cycle. It is apparent that the time required to machine a part, and hence the time during which the machine and the operator are engaged on the job, is much greater than the actual cutting time. These disadvantages are compounded when the operator has to make a number of similar parts from the same drawing. If their nature does not permit loading and clamping into jigs or fixtures, then inevitable errors of varying size, position and form will result.

In addition, many conventional machine tools have speeds and feeds governed by mechanical design features such as fixed-speed gearboxes. Thus, the choice of a feed or speed is a compromise depending on the gear ratios built into the machine tool. Optimum cutting conditions are rarely realised.

Many "automatic" machine tools have evolved over the years in an attempt to overcome some of the above limitations. Copy lathes, capstan lathes and turret lathes were early examples. Sequence control based on cams and later plug-board-operated pneumatic systems also made important contributions.

These approaches are characterised by extremely long set-up times by specialist setters. This meant that, once set up, the machines had to run for long periods and produce many thousands of parts to justify the long set-up times. It was common to over-produce whilst the machine was tooled up. This meant increased work in progress and working capital tied up in stock. Very often, production bottlenecks due to jobs queuing for certain machines disrupted production schedules.

By contrast, CNC machines offer complete control of all axes, under optimum cutting conditions. Extremely short **set-up times** are possible since standard tooling is all that is required. The need for jigs and fixtures is almost eliminated. Indeed, their presence can be an encumbrance to the flexible contouring facilities of CNC machines. Simple clamping arrangements are often all that is required.

Part programming is often carried out by specialist part programmers, away from the machine. The facility to prepare new jobs away from the machine means that the machine tool spends a greater proportion of its working time actually cutting metal.

Extremely good **accuracy** and **repeatability** of the components produced enables a greater uniformity of production. There are also attendant reductions in fitting costs, assembly costs, inspection costs and the elimination of scrap and re-work items. Moreover, once a job has been machined, the data that produced it (the part program) can be retained, saved and loaded back to produce identical parts at a later date. Figure 1/3 illustrates the comparison of machining components by CNC and conventional means.

The quality of the finished job is no longer under the control of the operator but under the control of a computer-run part program. This ultimately translates into lower costs per part and much-reduced **lead times**. Lead time is the term given to the time that elapses between an order being placed and the order being delivered.

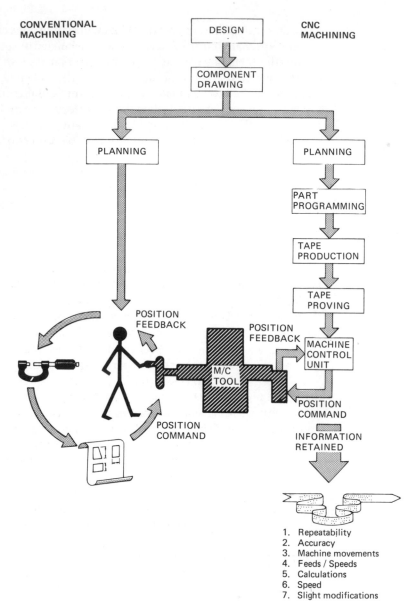

Fig. 1/3 Comparison of component production by conventional and CNC machining

CONVENTIONAL MACHINING

CNC MACHINING

DESIGN

COMPONENT DRAWING

PLANNING

PLANNING

PART PROGRAMMING

TAPE PRODUCTION

TAPE PROVING

POSITION FEEDBACK

POSITION FEEDBACK

M/C TOOL

MACHINE CONTROL UNIT

POSITION COMMAND

POSITION COMMAND

INFORMATION RETAINED

1. Repeatability
2. Accuracy
3. Machine movements
4. Feeds / Speeds
5. Calculations
6. Speed
7. Slight modifications

1.3/1 Profitable applications of CNC

For many of the above reasons, CNC finds greatest application in **small batch quantities** or **complex one-offs**. Traditionally, it is these jobs that prove the most difficult and costly to implement within a production environment. CNC is certainly not intended to compete with mass production techniques.

This should not be the only criterion on which the decision to adopt CNC should be based. There are numerous other applications to which CNC may be profitably applied. Examples include:

a) Where reliable, high-quality components are required.

b) Where operations, or set-ups, are numerous or costly.

c) When machine run time is disproportionately low, compared with set-up time.

d) When lead times do not permit conventional jig and tooling manufacture.

e) Where the part is so complex that quantity production involves the possibility of human error.

f) Where design changes, or individual variations, are required on a family of parts.

g) When inspection costs form a large proportion of total costs.

h) When tooling costs are significantly high, or where tool storage is a problem.

1.3/2 Adopting CNC

CNC provides for an automated approach to manufacturing. It gives greater scope to management's advance planning and more control of the actual manufacturing operation.

Flexibility is the hallmark of CNC machining. To change from one type of component to another requires only a minimum amount of set-up time and the provision of a part program. Modifications to existing components can be carried out swiftly and the part program updated almost instantly.

Tooling costs are drastically reduced since only standard, off-the-shelf tooling is normally all that is required. Part programs are not dependent on the size of the cutters used when the program was written. Most CNC machines are equipped with sophisticated cutter compensation and offset facilities which can accommodate variations in both the length and diameter of cutting tools.

The main reasons for adopting CNC can be listed as follows:

1 *CNC provides flexible automation, adaptable to many differing requirements, and changeovers from one job to another are rapid.*

2 *CNC produces components with repeatable accuracy in both dimensions and form.*

3 *The investment of time in producing part programs can be realised many times over since repeat orders require no additional work.*

4 *CNC places machine operation directly in the hands of management.*

5 *CNC makes short and medium production runs economical; therefore high volume production quantities are not required.*

6 *CNC offers reduced downtime, early production start-up, uniformity of production, minimum machining times, and less scrap and re-work.*

7 *CNC enables design changes in components to be quickly accommodated in minimum time with little disruption.*

8 *CNC leads to less work in progress which leads to quicker turnover of capital; hence less working capital is required.*

9 *CNC can reduce labour costs.*

10 *CNC is the first step towards flexible manufacturing, allowing the possibility of unmanned machining.*

1.3/3 Drawbacks of CNC

The limitations of adopting CNC are relatively few, but nonetheless will have far-reaching consequences for its adoption.

The biggest single limitation is probably that of the *initial high capital cost* of the equipment itself and its subsequent installation. This means that it is important that the machine is utilised (in actual cutting) as much as possible. This could impose a necessity for continuous shift working. A steady flow of work will thus be required in order to keep the machine(s) running. The costs become more prohibitive when the transition is made into flexible machining cells or, ultimately, flexible manufacturing systems. Such developments bring with them the need for considerable support systems concerning tool management, automatic work transport, etc., and the associated control technology to link them all together. Flexible manufacturing systems will be discussed in Chapter 8 (section 8.1).

The economics of CNC working are sensitive to machine reliability and continuous shift working will quickly exploit any such deficiencies. CNC equipment is also high technology equipment and will require maintenance of a different nature to that of conventional machine tools. The establishment of a resident, high-calibre *maintenance facility* is likely to be a prerequisite of adopting CNC.

Since most of the preparatory work for CNC operation is done away from the machine, *planned support facilities* will be essential. For example, part programming, tape preparation and tool pre-setting are fundamental services that must be considered.

CNC is a new discipline to those people not used to its operation. The approach to manufacture is different to that of conventional machining. For example, there is a shift in emphasis from manual skills in manipulating machine tool handwheels, to that of skills in planning and part programming. The existing workforce may need training or re-training, or new skills may have to be introduced into the organisation from outside. A *managed system of training* will need to be considered very carefully. It would be an intolerable situation to rely on a single operator for example. Under such a situation, if the operator became sick, then effectively the machine would become disabled. This quite obviously defeats many of the objectives of adopting CNC.

The introduction of CNC machines will inevitably lead to the loss of existing conventional machines and their operators. Whilst many new jobs may be created in the various support functions, jobs will undoubtedly be lost from the shop floor. This has far-reaching human connotations where people may be fearful for their jobs. Resistance to change may further hamper the introduction of CNC. Planned alternatives for those members of the workforce unable (for whatever reason) to make the transition into CNC operation must be carefully considered.

The introduction of CNC working into an organisation is a major undertaking. For many of the reasons outlined above, it is essential that careful pre-planning and feasibility studies, both in human and technical terms, are considered. The implications of getting it wrong can be serious and lasting.

1.4 CNC as a management control

1.4/0 Why a management control?

It is often quoted that CNC is not a machine control, but a management control. CNC is primarily an automatic machine, or process, control discipline. However, its intelligent adoption also greatly increases the potential control that management has over the production process. Management is concerned with the planning, organising and efficient usage of all factors and resources that are available. Most manufacturing organisations that fail do so, not because of an inability to manufacture, but because of ineffective management. Indeed, the responsibility for success or failure rests fairly and squarely with management.

In the context of considering CNC as a management control, it is useful to examine all the areas of manufacture on which CNC imposes an influence.

1.4/1 CNC and design

When a component has to be manufactured, it first has to be designed. It then has to be converted into a working drawing sufficient in detail for it to be produced. With the inception of CNC there are two considerations.

Traditionally, the way in which a part is designed is subject to the individual preferences of the designer. The adoption of CNC will immediately impose a certain influence on the design process. This, in turn, will influence the training of design personnel.

Designs will be originated for ease of production. In many instances an exercise known as **value analysis** can be carried out with the purpose of making production easier or cheaper. In the majority of cases, however, value analysis is usually carried out retrospectively on components that have already been designed and manufactured.

Detailed working drawings will be produced in accordance with accepted CNC conventions. Thus, it may be possible to include only salient dimensions and rely on the CNC system to bear the burden of repetitive calculations (see also Conversational Programming in Chapter 4, section 4.4/1.

In times of design change or modification, a simple edit at the machine may be all that is required. A current production tape can then be re-punched directly from the CNC machine control unit. Re-tooling, or re-jigging (and hence their re-design) can be alleviated, possibly eliminated.

The effect of the above is clear. Designers will be allowed to spend more time on the design process, rather than on the mere "housekeeping" tasks of updating drawings, etc.

The second consideration must be that the utilisation of computer power at the machine tool (via CNC) inevitably leads to the utilisation of computer power in other facets of the production process. Thus, the design process itself is being influenced by the application of computers, notably in computer graphics. The designers only function now is purely to design. Dimensioning, drawing production, filing, etc. can be carried out automatically by the system. Furthermore, designs created on the screen can be instantaneously converted into machining data, and transmitted directly to the memory of the CNC machine tool ready for machining.

1.4/2 CNC and process planning

The need for decision-making, on the shop floor, is largely eliminated as optimum speeds, feeds, travel times, tool selection, etc. are inherently built into the part program. Moreover, optimum machining conditions can be monitored and maintained by the CNC machine tool itself. Thus, optimum machining conditions can be specified and imposed more accurately by the adoption of CNC techniques.

1.4/3 CNC and production planning

The greatest application of CNC is in the small to medium batch product runs. Advantages are reaped in the very short changeover times that can be accomplished between jobs. Thus, lead times and production times are optimised, they become reliably predictable, and downtime is reduced to a minimum. This means that more accurate cost estimates can be generated, production runs can be planned more accurately, work in progress can be kept to a minimum, and customer service can be increased by specifying shorter delivery times and quicker response to modification requirements.

The need for extensive jigging and tooling can be considerably reduced, thus reducing their respective lead times. Finally, since the part programs can be produced, and proven, remote from the machine tool, more production time is available to allow the production areas to concentrate on producing.

1.4/4 CNC and quality control

Since the machining process is now automatic, better quality in terms of dimensional and geometrical accuracy is attained. Repeatability is increased, and this may enable inspection techniques to be re-designed. Because of the uniformity of the products, the processes of fitting and assembly can possibly be streamlined. Scrappage and re-work costs should be almost eliminated.

Inspection and measurement processes are themselves becoming subject to CNC control by the inception of specialist CNC measuring equipment.

The above points are valid and important. The decision to invest in CNC machine tools must not be taken on the basis of machining technology alone. It has far-reaching benefits for management if it is adopted intelligently. Of course, if management fails to acknowledge these additional benefits then the CNC installation will be less effective. The level of benefit is entirely dependent upon the user. It can range from a minimum where CNC is being used as a machine control, to a maximum where CNC is adopted as a management control.

1.4/5 Economic benefits to management

The economic benefits to management may be summarised thus:

1 *Better machine utilisation since the machine spends more of its time actually cutting metal.*

2 *The machine does not tire, require services such as toilets, food, drink, etc., require holidays, and will not go on strike.*

3 *Cycle times are considerably reduced.*

4 *The number of components produced is controlled by the machine and not by the operator; therefore delivery forecasts and production planning can be more accurately determined.*

5 *Small batch quantities (even one-offs) are as economical as large batches.*

6 *Savings on tooling, jigs and fixtures.*

7 *Accuracy is controlled by the machine; therefore better interchangeability (for spares), easier fitting and assembly, and reduced inspection are achieved.*

8 *Less scrappage and re-work.*

9 *Extended tool life since optimum cutting conditions are realised.*

10 *CNC machines may work unattended, in the dark and at other unsocial times. They can be integrated with other facets of the production process (planning, design, materials handling, etc.).*

11 *CNC machines can provide useful management information automatically.*

12 *Since machining time is more predictable, accurate costing data for financial control and estimating is more reliable.*

13 *The investment of time and skill in part programming can be saved and the benefits realised time and time again.*

14 *Adopting CNC puts the control of production back into the rightful hands of management.*

Questions 1

1 Briefly explain the meaning of Numerical Control.

2 When was numerical control introduced?

3 State the major differences between an NC machine tool and a CNC machine tool.

4 What factors have caused CNC machines to take over from NC machines?

5 Define the following terms relating to CNC operation: part program, block, sequence number, and machine control unit.

6 Why were the earlier NC machines referred to as "tape-controlled" machines?

7 What is meant by the term "battery back-up" in the context of a CNC machine tool and where would it be used?

8 Explain the difference between "single step" and "automatic" operation of NC and CNC machine tools, stating when each mode of operation would be used.

9 Why was it difficult to upgrade an NC machine but relatively simple to upgrade a CNC machine?

10 State *four* industrial applications of CNC.

11 What are Turning Centres and Machining Centres and why are they so called?

12 Define the term "lead time" and explain its importance.

13 Describe *six* advantages of employing CNC machining techniques over conventional machining techniques.

14 State *three* possible drawbacks of employing CNC machining techniques.
15 Under what circumstances can CNC best be employed?
16 Discuss the statement: "CNC is also a management control".
17 State *five* economic benefits that can be gained by adopting CNC.
18 In what ways could other manufacturing functions be affected by the adoption of CNC?
19 It is often stated that the introduction of CNC will cause a shift in emphasis from craft skills to technician skills. Discuss this point in relation to both present and future manpower requirements.
20 Comment on what you see as likely future developments in the field of manufacturing, due to the development of CNC techniques.

Machine tool considerations 2

2.1 Design differences between conventional and CNC machine tools

2.1/0 The need for design changes

Many design changes are required for dedicated CNC machine tools from their conventional manual counterparts. Some of the reasons are as follows.

Historically, machine tool designs have been kept as simple as possible consistent with the specifications they offered. This was largely to keep costs down. High-quality output was traditionally the responsibility of the skilled machine tool operator. The skilled craftsman could compensate for any deficiencies in the machine tool design. This was acceptable since labour costs were a reasonably low proportion of total manufacturing costs. Labour rates have now risen considerably and human factors can exert pressures that can render consistency in output somewhat unreliable. There is thus a growing tendency for machine tool manufacturers to build in features that will allow modern CNC machine tools to yield maximum output per man hour.

Set-up times and changeover times between jobs are drastically reduced under CNC working. As a consequence machine tools are spending more time actually cutting material. Because of the initial high investment of modern machine tools there is also a tendency to introduce multi-shift working and unmanned operation. The reliability of conventional designs becomes suspect. Machine tools originally designed for continuous operation over a single 8-hour shift are now, under CNC, required to work in excess of 20-hours under continuous operation.

Heavier demands are made on traditional designs. The cutting tool is brought to the cutting position much more rapidly than by manual means. The high percentage of cutting time results in faster wear on slideways, guide-ways, gears and leadscrews, etc. Optimum speeds and feeds, improved tooling and continuous path machining are subjecting the machines to high multi-directional forces not previously encountered on earlier machines.

Developments in modern cutting tool materials are demanding higher driving forces and greater rigidity in machine tool structures. Many modern cutting tool materials, whilst being very hard and wear resistant, are also very brittle. It is essential that all potential sources of vibration and backlash are designed out. Heavier structures, precision anti-friction bearings and high-quality components and assembly are thus desirable attributes.

The high rates of metal removal achieved with automatic operation have rendered traditional machine tool designs deficient in the rapid and effective removal of swarf from the machining area. In addition, a level of guarding is required with automatic machining that is not present on conventional machine tools.

Since the operator no longer has to control the functions of the machine via hand controls, traditional designs are no longer appropriate. It is now possible to combine functions into a single machine tool that previously only existed on separate machines. Modern machining centres, for example, are capable of milling, drilling, spotfacing, counterboring, threading, tapping and boring on up to five surfaces in a single set-up. Turning centres are equally versatile.

2.1/1 CNC machine tool design

All CNC machine tools are designed to fulfill three prime objectives:

a) To achieve and maintain accuracy.
b) To achieve and maintain repeatability.
c) To achieve and maintain reliability.

It is a common misconception that CNC machine tools are designed solely to reduce cycle times. Whilst they invariably do, it is more important that consistent quality of output is maintained and that downtime is kept to a minimum.

CNC machine tools enjoy considerable additions and improvements in four main design areas:

a) Structural details.
b) Mechanical details.
c) Control system facilities.
d) Auxiliary facilities.

As with most things, cost considerations will often force compromise. CNC machine tools are no exception. Discussion on the above points must be balanced by economic constraints that may influence design decisions. The old adage "you get what you pay for" may account for the wide variety of machine designs available.

2.1/2 Structural details

Structural errors are largely concerned with structural design and configuration.

STRUCTURAL CONFIGURATION   Structural configuration refers to the *external physical shape of the design and the relative positions of associated elements.* A horizontal machining centre, for example, has a different configuration to that of a vertical machining centre. Even within these types, different configurations may occur. A fully supported bed-type machining centre will

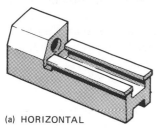

(a) HORIZONTAL

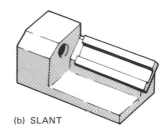

(b) SLANT

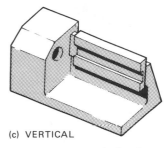

(c) VERTICAL

Fig. 2/1 Three turning centre configurations

Fig. 2/2 Machining centre configurations

be more accurate than a knee-type machine. Static and dynamic deflections of the machine structure will be more pronounced on a knee-type machine due to the increased overhang of the worktable. Hence, configuration can influence accuracy. Other factors may also be important. Three possible configurations for a turning centre are illustrated in Fig. 2/1.

Configuration (a) arguably provides greater support for the saddle and turret. Configurations (b) and (c), however, provide designs that allow swarf to fall naturally away from the cutting zone and allow the operator easier access to the workpiece and tooling.

The built-in accuracy of related machine tool elements can also be influenced by the machine configuration. The accuracy and rigidity afforded to the mounting of the machine tool spindle is a case in point. Alternative spindle head mountings for a horizontal and a vertical machining centre are illustrated in Fig. 2/2. On considerations of spindle overhang and resistance to twist of the structure, alternative (b) must be the preferred choice in both cases.

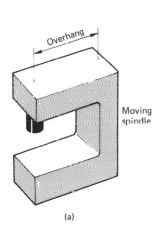

(a)

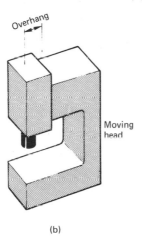

(b)

VERTICAL MACHINING CENTRE CONFIGURATIONS

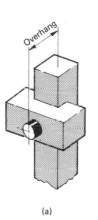

(a)

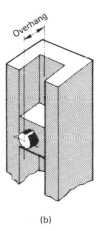

(b)

HORIZONTAL MACHINING CENTRE CONFIGURATIONS

STRUCTURAL DESIGN Structural design concerns the *physical construction of the structure*. Traditional machine tool beds, for example, are almost exclusively made from cast iron. This was because it is relatively easy (and less costly) to produce complex shapes; and cast iron has good damping properties and exhibits self-lubricating properties due to the presence of free graphite. Cast iron is still used as a basis for CNC machine tool construction although additional constituents are added to improve certain properties (nickel chrome Meehanite being one such material).

Accuracy is largely based on **rigidity**. Over-proportioned guideways, slideways and spindles are characteristic of modern CNC machine tools. This is to resist the high torsional forces and heavy-duty cutting conditions imposed by high machine utilisation. As a consequence, thicker sections will be found on most modern cast structures.

Steel welded structures have also found a place in machine tool construction. Boxed structures offer considerable rigidity, inherent balance and symmetry, and considerable flexibility in the mounting of ancillary components. Because of the increased strength of steel and the flexibility to fabricate structurally efficient designs, steel structures tend to be of lighter construction. The use of intermittent welds helps with vibration resistance and damping.

In pursuit of greater accuracy, hardened slideways are often hand-scraped. In addition to attaining high precision and a smooth low-friction surface, oil retention is considerably increased, a factor that does much to minimise wear. Where flat bearing surfaces are used, low-friction PTFE (polytetrafluoroethylene) coatings can be used. In many cases, continuous, forced lubrication systems operate to reduce friction and wear.

A major source of inaccuracy within machine tool structures is **thermal distortion**. Heat generation may occur from many sources, both internally within the machine tool and externally in the machine shop environment. Solar heat falling onto machine tools through windows and body heat from personnel are common external sources of heat. Sunlight may be masked out using curtains or blinds and the effects of body heat may be combatted by installing temperature control or air conditioning facilities.

Sources of heat from within the machine tool include the headstock gears, motors, bearings, transducers, transmission elements, friction, swarf and the cutting operation itself. Such sources of heat can cause the machine tool structure to distort. Structural design can help reduce thermal distortions. Symmetrical structures will tend to distribute any effects evenly. The mounting of the headstock, motors and other sources of heat centrally within structures, will have the effect of distributing loads and heat evenly through the structure. Refrigerated oil and coolant systems that maintain lubricating oil and coolant at ambient temperatures are sometimes used.

Many manufacturers recommend warming-up times for their machine tools at the beginning of the day, or at the start of a new shift. This allows thermal equilibrium to be established within the machine tool structure before machining takes place. Indeed, the best way to combat the effects of thermal distortion may be to keep the machine tool working constantly.

2.1/3 Mechanical details

Transmission of motive power in driving spindle and worktable elements are the chief sources of mechanical inaccuracy.

Consider the drive arrangements required to move the worktable. From the motor (other than for direct drive applications), transmission will pass through a number of stages.

GEAR BOX/BELT DRIVE STAGE   Gearbox errors include backlash and friction due to preloading. To overcome these errors, purpose-designed single-ratio gearboxes with light preload are used. A **single-ratio gearbox** is one that only has two gears in mesh at any one time. Accumulation of backlash between many gears in a train is thus eliminated. Where belts are employed it is more favourable to use **toothed belts**. These are mechanically positive and are not subject to slip. Belts, however, are unsuitable for vertical drives (because the machine could be rendered unsafe due to breakage), and high power applications (because of the tendency to stretch).

MOTION TRANSMISSION STAGE   Approximately 90% of machine tool applications utilise leadscrew drives. The use of Acme-form screw threads (found on most conventional machines) is not suitable for CNC applications for the following reasons. They exhibit high friction/high wear characteristics, have poor power efficiency, and produce high lost motion due to excessive backlash. In addition, they can only be used at relatively low rotational speeds. They are not accurate or repeatable enough for the accuracy demanded by CNC.

They have been almost universally superseded, in NC and CNC applications, by the now familiar **recirculating ball leadscrew**. This effectively replaces sliding friction with rolling friction, rather like replacing a plain bronze bearing with a ball race anti-friction bearing. Both the leadscrew and the nut have a precision ground form into which ball bearings are allowed to run. The form is so designed that contact only takes place on opposing faces of the assembly. The geometry of the ground thread form can either be semi-circular or a "gothic arc" (ogival) form. Enough ball bearings are inserted to completely support the screw/nut assembly. Internal or external ball tracks allow the balls to continuously circulate as the screw rotates in the same manner as within a ball race. Such leadscrews are very efficient and a load carried in a vertical plane is quite capable of sustaining a downward movement under its own weight. The rigidity of a drive system, and the positioning precision, can be increased by pre-loading the nut assembly. This is achieved by using two nuts and mounting them such that a pre-load exists between them. They may be pre-loaded in tension or compression. Pre-loading tends to increase wear and increase the torque required to drive the screw. A gothic arc form is the preferred design. Both semi-circular and gothic arc forms are illustrated in Fig. 2/3.

For extremely high precision machine tool applications, CNC coordinate measuring machines and high-quality industrial robots, **a planetary roller leadscrew** can be employed. The planetary roller screw employs precision threaded rollers which engage with a precision ground leadscrew. As the leadscrew is rotated, the rollers act like a planetary gearbox by rotating around the leadscrew within the nut assembly. The rollers have gear teeth cut into each end. These gears mesh with the internal teeth of a gear ring within the nut assembly. This prevents axial movement of the rollers but allows linear motion to be transmitted through the assembly. High precision movement can be achieved since the precision leadscrew can be ground with a multi-start thread. The multiple contact points achieved by the precision threaded rollers offer heavy load-carrying capabilities. Longer working life is achieved since the multiple contact points evenly distribute the applied loads. The threaded rollers are continuously in contact with both the nut and the leadscrew. There are no "loose" balls to recirculate and thus higher rotational speeds can be achieved. Recirculating ball leadscrews are shown in Fig. 2/4 and a planetary roller screw is in Fig. 2/5.

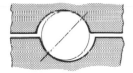

SEMI-CIRCULAR

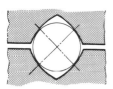

GOTHIC ARC

Fig. 2/3 Recirculating ball leadscrew forms

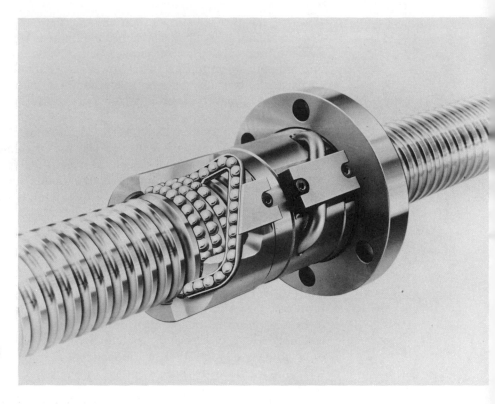

Fig.2/4a Recirculating ball screw: external return
[*PGM Ballscrews*]

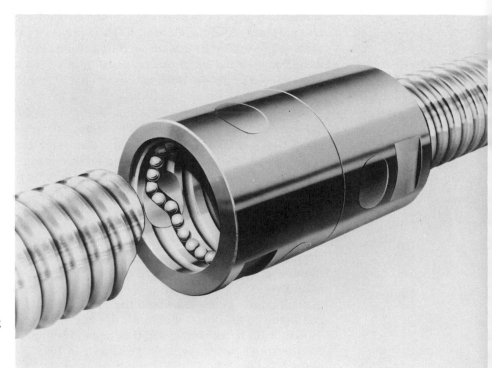

Fig. 2/4b Recirculating ball screw: internal return
[*PGM Ballscrews*]

Fig. 2/5 Planetary roller screw [*ESE Ltd*]

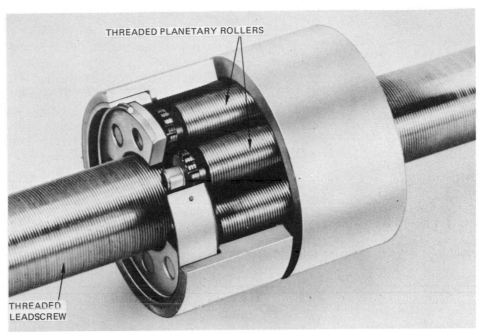

THREADED PLANETARY ROLLERS

THREADED
LEADSCREW

Where table travel exceeds two-thirds of a metre, screw drives become less attractive. High-inertia torsional losses and the tendency for the screw to sag under its own weight are typical problems. For large displacement applications, *worm and rack* or *rack and pinion* drives may be used. These systems, however, tend to be more expensive to implement.

Hydraulic table-positioning systems have been used (via hydraulic rams) with some success. They offer relatively high performance for a relatively low cost. They are restricted to the smaller applications where the product (mass × stroke) for the application concerned is relatively small. Where large displacements dictate long piston strokes, the large volumes of hydraulic fluid required can be subject to slight elastic compression which can result in positional inaccuracy. Dirt, noise and leakages are additional problems.

SLIDEWAY STAGE   Conventional machine tools almost universally rely on direct contact between the slideway and other moving elements. Relative movement between the surfaces is relatively slow (due to hand operation), and machine utilisation is fairly low. This is an adequate (and reasonably cheap) state of affairs. Hardened, ground and lubricated slideways together with the self-lubricating properties of the cast iron used, reduce wear to reasonably acceptable limits. CNC machines are much more demanding on slideways. Low inertia and high acceleration and deceleration rates are required under automatic control. Responsive action without stick-slip effects is essential for positioning accuracy. **Stick-slip** is a condition that occurs when small movements between two lubricated elements are required. The effect of the lubricating medium tends to cause the mating elements to "stick" to each other and resist motion. A resultant "jerky" action is produced as the two elements tend to "stick" and then "slip" during their relative movement. Wear resistance, high stiffness and good damping qualities are also desirable. Low-friction bearing surfaces are thus essential in CNC machine tool applications. In addition to aiding the above points, low-friction bearings also reduce the amount of power required to bring about worktable movement.

The approach (rather like that of the leadscrew) is to replace sliding friction (as a result of direct contact) with rolling friction by the use of anti-friction ball and roller bearings. This can be achieved in a variety of ways. A **linear ball bush** utilises the principle of recirculating balls within the design of a bush-type bearing. They are designed to run along precision ground shafts. They offer frictionless movement over varying stroke lengths with high linear precision. Up to six recirculating ball orbits, running the complete length of the bush, are provided. Where appropriate, rotary movement around the shaft can also be achieved. Linear ball bushes are illustrated in Fig. 2/6.

Where movement along a flat plane is required **recirculating roller bearings** can be employed. These are self-contained, precision units that employ recirculating precision rollers. Their appearance resembles that of a skate. The precision rollers can withstand high loads and, since they recirculate, offer unlimited travel. Accurate machine alignment and motion is achieved by using them in pre-loaded pairs. The units can be mounted horizontally for load-carrying applications such as machine tool tables. They can also be mounted "sideways on" to provide support, guidance and motion for vertical machine tool elements. Used in this way the units resist any sideways movement of a machine tool worktable. The self-contained units (available in various sizes) provide for ease of assembly onto hardened machine surfaces or hardened inserts. The condition of stick-slip is eliminated and smooth and precise movement is achieved. A linear roller bearing is illustrated in Fig. 2/7.

Fig. 2/6 Linear ball bushes [*ESE Ltd*]

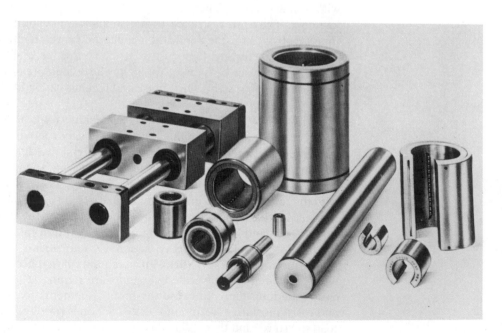

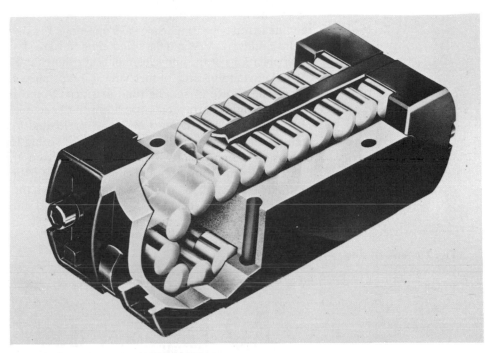

Fig. 2/7 Linear roller bearing [*ESE Ltd*]

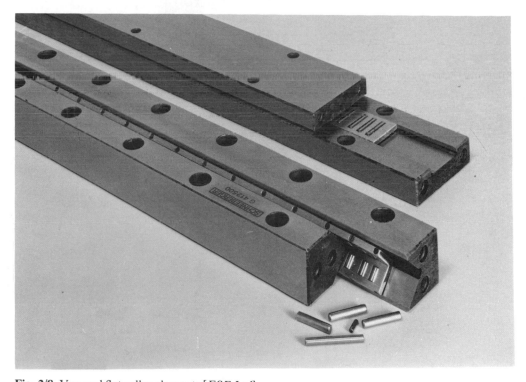

Fig. 2/8 Vee and flat roller elements [*ESE Ltd*]

Frictionless linear movement can also be provided by a **vee and flat roller** arrangement. This arrangement utilises precision rollers in a similar way to the linear roller bearing. Sideways motion is resisted by the structural design of the guideway. Vee and flat roller elements are shown in Fig. 2/8.

Applications of the above elements are shown in Fig. 2/9.

The outstanding area of mechanical importance concerns the *main spindle drive*. The choice of spindle (and axis drive) motors is discussed in Chapter 3. The mounting arrangements within the headstock will typically consist of high-quality angular contact bearings, preloaded for stiffness and rigidity at high speeds. Rear spindle bearings may be spring-loaded to take up thermal expansion. High-quality hardened, ground and balanced drive gears will reduce friction. Vibration and noise may be further reduced by employing toothed belt final drive arrangements. An automatic filtered pressure lubrication supply may be included. Pressure and level sensors will prevent damage in cases where lubrication supply is interrupted.

Fig. 2/9 Anti-friction slideway system examples

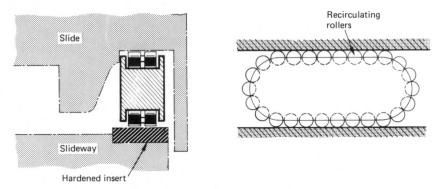

RECIRCULATING ROLLER SLIDE

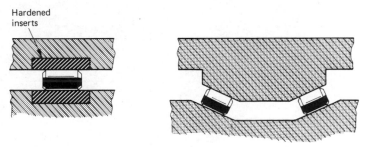

VEE AND FLAT ROLLER SLIDE

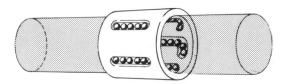

LINEAR BALL BUSH

2.1/4 Control system facilities

Traditional manual operation of the CNC machine tool is not usually possible. Indeed one of the first features apparent on most CNC machine tools is that there are no longer any handwheels to be seen. Control of all machine functions has been totally transferred to the computerised control system (from which CNC derives its name). This generally takes the form of a control unit situated within convenient reach of the operator. The control unit may be housed within the machine structure itself or mounted on a swinging arm arrangement allowing it to be adjusted according to the position of the operator. Ergonomically this is more acceptable to the operator than having to reach for handles, handwheels, levers and switches.

The facilities offered by the control unit may be considered under three broad headings. Firstly, those facilities which will indicate the current status and position of the various machine tool features. Secondly, those facilities which allow manual or semi-manual control, of the machine tool movements, and thirdly those facilities that enable the machine tool to be programmed. Such facilities will be present on all CNC machine tools to a greater or lesser extent. The common features likely to be encountered will be discussed.

STATUS INDICATORS   In order to control machine tool movements they must first of all be identified. There is a standard system of axis designation that will be explained in the following section. The common machining centres will have X, Y and Z axes of control and the common turning centres will have X and Z axes of control.

All CNC machines will have some means of indicating the current position or movement of the **controlled axes**. Some machines may have simple digital readouts. These may be dedicated to each axis or there may be just a single display which must be switched to indicate the desired axis. More expensive machines employ a VDU (visual display unit). This is a screen resembling that of a television. Such screens are very versatile and may be used for viewing and searching part programs, examining tool information, showing graphical plots of the tool path via computer graphics, and for diagnostic purposes. In the majority of cases the same facilities will be used to display the part program. This may be displayed as machining commences, or for viewing, searching and editing operations.

Axis movements may be shown in one of two ways, selectable by the operator. The absolute position from the machine datum may be continuously displayed. If desired, the incremental movement may be displayed. This is the movement of the slide from the last point visited.

Other indicators may included *power consumed* and *spindle speed*. Power consumed is a good indicator of the efficiency at which the machine is operating. There will be the usual array of warning and indicator lamps.

MANUAL/SEMI-MANUAL CONTROLS   There will be a need to operate the machine tool other than under program control. For example, when loading, unloading or setting work, establishing machine datums, loading or changing tools, trying out and proving part programs, for cleaning and maintenance, etc., etc. There will obviously be some means of effecting axis movement.

The operator will usually be able to select whether such movement will take place under rapid traverse or feed. Most machines also have access to a **jog mode**. This may also be termed *inch* or *step mode*. In jog mode, one depression of the appropriate axis control button will move the machine slide by a predetermined amount of movement. The operator will select the amount of movement required. Typical values might be 1-unit, 0.1-unit or 0.01-unit. This facility is primarily used to establish datums and "pick-up" on certain component features. This is what is understood to be manual operation.

When a part program is loaded into the machine it is desirable to execute it step by step, a block at a time. This is called **single-step operation** and is the equivalent of semi-manual or semi-automatic operation. In single-step mode, a single-key depression will execute a single part program block of information and stop.

Fully automatic operation may of course be selected to execute complete part programs. There are a number of **manual override** facilities, at the disposal of the operator, which may be invoked whilst the part program is being executed. The two main facilities are *feed override* and *spindle speed override*. Typical arrangements allow the programmed values to be modified down to rest and up to 120% of programmed value. This allows the operator to fine-tune the cutting operation in the light of local machining conditions.

There will be facilities to stop and re-start programmed operation, and the provision of emergency stop facilities.

PROGRAMMING CONTROLS   There will be facilities to allow part programs to be installed into the memory of the CNC control unit. The mode of data input will be selectable, depending on those available. The most common mode will be **manual data input** (MDI). In this mode, the operator is allowed to enter and edit part programs, a block (instruction) at a time. This will be accomplished by a keypad offering the required facilities. Other input modes that may be available include magnetic disc or tape, punched tape (from a tape reader or teletype device), or direct from a host computer (via an RS232C interface). These means of data input are discussed in Chapter 4, section 4.4.

Other programming controls relate to the entry and editing of values that effect tooling. These will be discussed in Chapter 7, section 7.4.

Finally, many control units are equipped with a battery back-up facility. This enables the contents of the control unit memory to be retained even when the main source of power is removed. This is extremely useful in cases of power failure or where the same job is being run day after day. The part program (or programs) need be installed into memory once only. Reloading at the beginning of each working day is eliminated, thus saving time that can be more usefully employed cutting metal.

2.1/5 Auxiliary facilities

Perhaps the most obvious outward feature that distinguishes the CNC turning centre from its manual counterpart, the lathe, is the appearance of extensive **guarding**. Certainly most turning centres are almost totally enclosed by an envelope of elaborate guarding. This is to afford protection from the large amounts of swarf and coolant produced in the cutting zone. The high spindle speeds involved create high forces on components, machine tools and cutting tools alike. The consequences of tool or component breakage or a serious collision necessitate such comprehensive measures. Machine tool guards will often be interlocked with the control system via limit switches activated upon closure of the guards. If the guards are not fully in position then machining is prevented from taking place. Machining centres do not generally permit such a high level of guarding and it is often accomplished at a more local level.

Under no circumstances should machine tools be used without adequate guarding and eye protection. Interlocking devices on guards and other safety-related equipment should be maintained in fully operational order.
Refuse to operate machine tools on which such devices have been disconnected or where adequate protection is not available.

Most CNC machine tools will be fitted with **automatic lubrication** facilities. This is to minimise wear due to the high levels of continuous operation and to extend the life of machine tools requiring high capital investment.

Some of the more elaborate CNC machine tools have **automatic tool-changing** (ATC) facilities. Such features are hardly ever implemented on conventional machines. Modern turning centres are also offering automatic chuck jaw changing. This allows for different components to be machined under program control giving scope for unmanned operation. *Automatic component-loading* facilities (or the facilities to upgrade to such) are being implemented on newer CNC machines. Such things as component feeders or automatic pallet changers (APC), facilitating dual set-ups, offer alternatives to robots at a local level. Facilities for adding adaptive control (see Chapter 8) may also be available.

Most levels of CNC control offer inbuilt **diagnostic** capability. Stored programs can execute checks to verify the operation of various components within the control system. This speeds up the diagnostic process in isolating errors and effects speedier maintenance.

Since the control system and its attendant features are computerised, many of the facilities are brought about by software control: that is, by stored computer programs within the CNC unit itself. CNC machine tools offer considerable scope for updating machine facilities. It is a far simpler matter to alter a stored program (usually by changing a printed circuit board or chip), than it is to modify or adapt mechanical details.

2.2 Classification of CNC machine tools

2.2 / 0 Types of CNC machine tool

Mention has already been made of machining and turning centres. They are so called because they are capable of performing (automatically) many operations, in a number of planes, in a single set-up. Very often, automatic tool-changing facilities and automated swarf-removal systems are also incorporated. Optional features can include automated component loading and adaptive control facilities. The variety of operations performed by such machine tools would require the use of many different types of conventional manually operated machine tools. CNC machine tools are the building blocks of Flexible Manufacturing Cells and Systems (see Chapter 8), and may be integrated with other functions in the total production process.

In the context of the following discussion the term **machining centre** will encompass all varieties of CNC milling machine and the term **turning centre** will, similarly, encompass all varieties of CNC lathe.

The most common CNC metal cutting machine tools fall into two distinct categories:

1 *Those that employ rotating tools and cutters whilst maintaining the work stationary.*
2 *Those that require the work to rotate whilst maintaining fixed tools.*

The first category comprises both horizontal and vertical machining centres. The second category is made up largely of turning centres. More recent turning centre designs allow the facility for power-driven tooling (drills, etc.) within the tool turret. These machine tools can thus exhibit both rotating tools and rotating components. Where rotating turret tools can be operated independently of the main spindle, small milling-type operations can be accomplished. This further enhances the type of work that can be accommodated on such turning centres and displays even greater flexibility.

An important recent addition to the range of CNC machine tool types is a machine known as the **CNC mill-turning centre**. As its name suggests it is able to carry out a number of both milling and turning related functions within a single machine tool. The outward appearance is like that of a turning centre. It has two separately programmable slides, based on the slant bed configuration. Both slides have tool turrets but the upper slide has the facility for driven tools. The upper turret may also be moved (under program control) radially, thus giving it the action of a compound slide. Both slides may be used simultaneously. Since the upper slide houses driven tools, it has the capacity to accommodate small milling cutters which may perform milling operations on the component held in the main chuck.

Also on the mill-turning centre a number of fixed tools can be provided, opposite the upper turret. This can enable the very versatile facility of performing machining operations on the rear "part off" face of components. The component will be finished-turned. A portable chuck, situated in one of the turret positions of the upper slide, can be programmed to rotate and simultaneously grip the component in the main spindle chuck. The component may then be parted off leaving it held in the auxiliary chuck. Under program

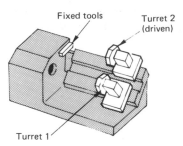

Fig. 2/10 A Mill-Turning centre configuration

Fixed tools   Turret 2 (driven)

Turret 1

control, the upper turret can manoeuvre the component to be faced, drilled, tapped or turned by the stationary tools opposite the upper slide. This means that a component may be finished-turned/machined without interruption by the operator (or other manipulating devices) having to reverse or remove the workpiece.

The configuration of a mill-turning centre is illustrated in Fig. 2/10.

2.2/1 Axis identification

Controlled axes on CNC machine tools are identified according to established standards. BS3635 Part 1:1972 illustrates the axis classifications of twenty-five CNC-related machine types. This section will deal with the commonly controlled axes on CNC machining and turning centres.

The basis of axis classification is the 3-dimensional Cartesian coordinate system. This is the system employed for graphical plotting in mathematics. In machine tool terms the axes correspond to **longitudinal**, **transverse** and **vertical planes** of movement. The three dimensions of movement are identified by the upper case letters X, Y and Z. It is also necessary to be able to identify the direction of movement along each of the controlled axes. Direction of movement is specified as being either "plus" (+) or "minus" (−) from an established machine datum, again according to established standards.

The main axes of movement are identified below.

The Z-axis The Z-axis of motion is always parallel to the main spindle of the machine. It does not matter whether the spindle carries a rotating tool or a rotating workpiece. On vertical machining centres, the Z-axis will be vertical. On horizontal machining centres and CNC turning centres, the Z-axis will be horizontal.

Positive Z movement (+Z) is in the direction that increases (or would increase) the distance between the workpiece and the tool. On vertical machining centres, +Z movement is always away (upward) from the machine worktable. On horizontal machining centres and turning centres, the direction of +Z motion is always away from the spindle.

The X-axis The X-axis is always horizontal, it is always parallel to the workholding surface, and it is always at right angles to the Z-axis.

If the Z-axis is vertical (as on vertical machining centres), positive X-axis movement (+X) is identified as being to the right, when viewed from the spindle looking towards its supporting column.

If the Z-axis is horizontal (as on horizontal machining centres and turning centres), positive X-axis motion is to the right, when viewed from the spindle towards the workpiece.

Fig. 2/11a Three-axis cartesian coordinate system

Fig. 2/11b Primary axes of movement for common CNC machine tools

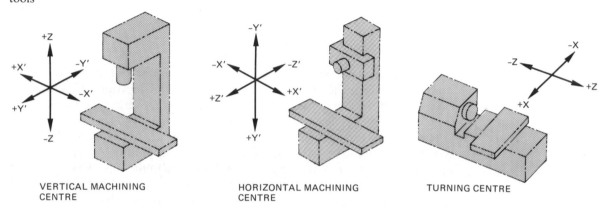

VERTICAL MACHINING CENTRE

HORIZONTAL MACHINING CENTRE

TURNING CENTRE

The Y-axis The Y-axis is always at right angles to both the Z and X axes. A CNC turning centre has only two major axes of movement at right angles to each other. CNC turning centres do not have a Y-axis of motion.

Positive Y-movement (+Y) is always such as to complete the standard 3-axis coordinate system. The +Z and +X directions will effectively determine the direction of +Y when superimposed onto a graphical representation of the 3-axis coordinate system. This is illustrated in Fig. 2/11a.

The primary axes of movement for the common CNC metal cutting machine tools are identified in Fig. 2/11b.

2.2/2 Additional axes of movement

It is common for CNC machining and turning centres to have additional linear axes of movement, often in parallel with the three primary axes. For example, a turning centre may have a horizontal saddle movement and a horizontal turret movement both operating in the X-axis. Obviously the control system must be able to distinguish one from the other in order to command appropriate movement of the correct element.

In general, where there is more than one moving element in the same axis, one is identified as being the primary movement, and is allocated the primary axis designation X, Y or Z. It will usually be the element nearest to the spindle in the axis concerned. Secondary movements in the same axis

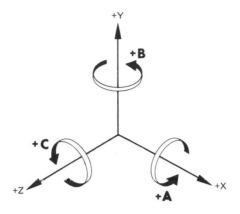

Fig. 2/12 Identification of rotary axes of movement

are then designated the upper case letters U, V and W, corresponding to secondary motion in the X, Y and Z axes respectively.

It is also possible for rotary movements to be provided as part of the original machine in the form of built-in rotary tables. This should not be confused with rotating tool turrets found on many turning centres. A rotary tool turret is not an axis movement, it is merely a tool selection device.

Where rotary axis movements are provided they are identified by the upper case letters A, B and C which correspond to rotary movements about the X, Y and Z axes respectively. Clockwise rotation is designated positive movement and counter-clockwise rotation as negative movement. Positive (clockwise) rotation is identified by looking in the +X, +Y and +Z directions respectively. The amount of rotary motion will be stated in degrees.

This convention is illustrated in Fig. 2/12.

2.2/3 Reversed direction for moving workpieces

The above designations are accepted standards and as such apply to all CNC metal cutting machine tools. Part program commands requesting axis motion will reference the axes and directions of motion according to the above designations, i.e. +X256.25 or −Z5∅ as appropriate. This is made on the assumption that it is the tool that is being moved. A moment's thought will confirm that, if a machine tool moves the workpiece instead of the tool, it must respond to the motion command in the opposite direction to that defined above.

This anomaly is taken into account by the machine tool and control system designers. In all cases the part program commanding axis movements should be written on the assumption that it is the tool that moves. It does not matter that it may, in practice, be the workpiece that actually moves.

This can, however, cause some confusion in indicating direction of axis movement on technical diagrams of machine tools. In cases where it is the workpiece that moves (and table movement is in the opposite direction to that specified above), positive movement of the *axis* is shown by a primed (or dashed) letter thus: +X′. This is read ''X dash'' or ''X prime''. A primed letter indicates that worktable movement is required in this direction in order to bring about relative tool movement consistent with the designated positive axis directions. (Fig. 2/11 *b* will confirm the notation.)

2.3   Swarf control

2.3/0   The need for swarf control

In parallel with substantial developments in CNC machine tool and control system design, there have been equal advances in cutting tool technology. Revolutionary new cutting tool materials and designs have been developed in an attempt to keep pace with continuing machine developments. The result has been a substantial increase in the cutting speed at which many materials can be machined.

CNC machines are now capable of achieving optimum cutting conditions under continuous (and often unattended) operation. This has meant a dramatic increase in the volume of swarf that can be removed in a short period of time. Even modest machine tools are capable of exceeding metal removal rates in excess of $500 \, \text{cm}^3/\text{min}$. At this rate of metal removal it will take just over 4 hours to accumulate 1 tonne of mild steel swarf. Unless swarf is quickly and efficiently removed from the cutting zone it can impede the cutting process and degrade the quality of the finished product. If it is not disposed of from the machine tool area itself it can considerably hamper access when attention to tooling or the component is required. Auxiliary functions involving automatic tool changing or automated component loading may also be affected by large accumulations of swarf.

2.3/1   Swarf control systems

Traditionally, swarf removal is seen as an irritation that the operator has to live with. Not a great deal of attention is given to it. It is nearly always carried out by hand and labourers are often employed to service a number of machine tools. With the automated nature of CNC machining, the high rates of material removal and an emphasis tending towards unmanned manufacture, new attention has to be focused on the very real problem of swarf removal and control.

In many cases, swarf control equipment will be provided to support existing CNC installations. In the case of acquiring new CNC machine tools, swarf control needs to be considered as being important in its own right and not merely as a necessary appendage to the manufacturing process. Many CNC machine tool manufacturers have recognised the problem and are now designing in swarf removal and collection facilities as standard equipment.

Swarf control often represents an additional capital burden on installation, maintenance and running costs. This should be realised when considering CNC operation. It is an oncost that must be reflected within the total running costs of the CNC installation, for example when tendering or estimating for contract work.

Swarf control must accommodate both removal of troublesome swarf from around cutters and the cutting zone, and the disposal of swarf from the machine tool area itself.

2.3/2 Swarf removal from the cutting zone

As we have seen, the removal of swarf from the cutting zone can be helped considerably by the design configuration of the machine tool itself. Horizontal machining centres are at a distinct advantage to vertical machining centres in that swarf falls away naturally from the cutting area. Chute arrangements can be provided to guide swarf away on vertical machining centres. Slant-bed and vertical-bed turning centres enjoy the same advantage over the traditional horizontal-bed configuration of the traditional lathe.

Swarf removal by gravity is not usually adequate in itself and it must be supplemented by additional means. Multiple coolant jets arranged around the cutting tool are sufficiently effective to rid the tool of accumulated swarf. Flood coolant is often supplied under pressure for this purpose. Coolant may also be effective in ridding the worktable area from swarf during cutting. Where automated inspection or probing techniques may follow the completion of a component, it is common to include a "coolant wash" stage within the program. This is to wash the component free of swarf prior to the inspection stage.

There are many periods when swarf removal by coolant is not possible. At the end of the cutting cycle when new components have to be loaded, during programmed breaks for inspection, or clamp position changes, the spindle and coolant flow is usually stopped. Swarf removal is still very necessary at these times. It would be unthinkable to load a component onto a dirty, swarf-laden worktable. In these cases, compressed air jets are sometimes used as an alternative to manual swarf clearance. Robots can be used to pick up air lines and direct air jets to perform this function.

> *Users should be very sternly warned of the dangers of misusing compressed air. Under NO CIRCUMSTANCES should compressed air jets be directed at the human body. The consequences of air embolism within the bloodstream can prove fatal.*

Another approach that can be taken is to employ purpose-designed swarf collection devices. These operate on the principle of the domestic vacuum cleaner. They may be powered by compressed air or by conventional electric motors. They have the advantage that swarf removal and disposal can be achieved in one operation. A network of pipes and pumps can extract the swarf to a central collecting point.

2.3/3 Swarf disposal from the machine tool

Swarf disposal from the machine tool area is usually accomplished by mechanical means. Continuously operating linear transport or screw conveyors can be arranged within, or external to, the machine tool. Swarf falling onto the conveyors is immediately transported away and does not have time to accumulate. Typical systems are illustrated in Fig. 2/13.

The destination of the swarf may be to individual collecting bins or to a central collection point. Where individual collecting bins are used, it is

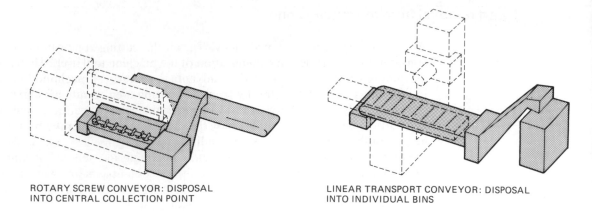

ROTARY SCREW CONVEYOR: DISPOSAL
INTO CENTRAL COLLECTION POINT

LINEAR TRANSPORT CONVEYOR: DISPOSAL
INTO INDIVIDUAL BINS

Fig. 2/13 Swarf disposal systems

a simple matter to keep swarf from different materials separate. This may be important for the reclamation of valuable materials.

Where a central collection point is used, swarf can be transported by manual means (drums or skips) or automatically via conveyor systems operating in channels provided in the floor. Alternatively, conveyor systems above floor height may be preferred. This is a more permanent and less flexible alternative as far as factory or shop floor layout is concerned.

The ideal system may prove to be a dual system incorporating both methods. Swarf can be automatically disposed of into individual skips direct from the machine tool. Automatically guided vehicles (loaded by robots?) can then transport this to a central collection point for swarf processing operations.

Swarf can inflict painful injuries if mishandled. If it is necessary for the operator to remove swarf, gloved hands should be used. Under no circumstances should swarf be handled in the cutting zone with either the spindle rotating or the machine slides in motion. Eye protection must be worn at all times.

Questions 2

1 Why do CNC machines require design changes over their manual counter-parts?
2 State *four* areas where design changes are required.
3 Discuss the factors that affect the structural design of a CNC machine tool.
4 Why has "rolling friction" replaced "sliding friction" in the design of machine tool slideways?
5 Define the following terms in the context of CNC machine tools: jog mode, feed and speed override, manual data input, and recirculating ball leadscrew.
6 Discuss the relative advantages and disadvantages of direct drive, gear box drive and belt drive transmission systems.
7 Explain why CNC machine tools have to be fitted with comprehensive guarding and interlocking devices.
8 What is a Mill-Turning Centre and why is it so named?
9 Why are the Acme-form leadscrews found on most conventional machine tools not suitable for CNC machine tools?
10 State *one* advantage and *one* disadvantage of a recirculating ball lead-screw.
11 Why is it that optimum cutting conditions can be achieved on CNC machine tools but not on conventional machine tools?
12 State *two* means by which wear resistance is increased on the slideways of CNC machine tools.
13 Make neat sketches of a typical turning centre, a horizontal and a vertical machining centre and indicate the primary X, Y and Z axes of motion. Explain, by reference to your sketches, what is meant by positive and negative movement.
14 Explain, with the aid of sketches, where the auxiliary axes of motion A, B, C and W would be found.
15 What is meant by the term "swarf control" and why is it important when dealing with CNC machine tools?
16 State *two* means of removing swarf from: *a*) the cutting zone and *b*) the machine tool.
17 On some CNC machine tools it is the spindle that moves and on others it is the worktable that moves in response to programmed movement commands. How is this taken into account when producing a part program?
18 State *three* sources of heat that could cause the machine tool structure to distort and suggest a possible remedy for each.
19 State *three* sources of vibration within a machine tool and how these have been reduced in the design of CNC machine tools.
20 Explain, giving examples, the difference between "static" and "dynamic" loading of a machine tool structure.

A flexible manufacturing system *(Courtesy: Cincinnati Milacron)*

Control considerations   **3**

3.1   Open and closed loop control

3.1/0   Introduction to automatic control

Proper control of the machine tool motions is an essential requirement of any CNC machining application. Without it we would still be reliant on the manual skills of the conventional machine tool operator. But how is this automatic control achieved? Consider first some fundamental factors relating to control systems in general.

For our purposes, a **control system** may be defined as:

One or more interconnected devices which work together to automatically maintain or alter the condition of the machine tool in a prescribed manner.

Such a system may be mechanical, electrical, electronic, hydraulic or pneumatic. In practice, many control systems are combinations of these and are termed **hybrid systems**.

In theory, an *input signal* is generated in response to an inputted program command. This produces an *output signal* which turns a motor, which then moves the machine tool slide. In practice, however, to achieve this satisfactorily can be a complex problem.

One important distinction that must be made in relation to control systems, is between *open loop* and *closed loop* operation. Consider the following example.

Suppose we have a furnace heated by an electrical heating element and controlled by a dial graduated in degrees C. We require a furnace temperature of 100°C and set the dial accordingly. This represents an input command and generates an input signal (voltage) to the heating element. In turn this input signal produces an output (electric current, and hence heat) which controls the final temperature. The temperature is known as the *controlled quantity* and will eventually stabilise at a *steady state* value. If the dial has been calibrated correctly, this steady state value will be 100°C.

On the face of it we have a fairly simple control system for furnace temperature as illustrated in Fig. 3/1.

However, consider what happens if the door of the furnace were left open. The temperature obviously drops, yet the input signal (dial reading) and the

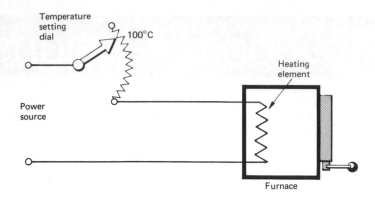

Fig. 3/1 Simple furnace temperature control

Temperature setting dial

100°C

Power source

Heating element

Furnace

output signal (current) behave as though they were delivering the requested 100°C. Similarly, if we were to place a red hot casting into the furnace and close the door, the temperature would rise considerably. Again the dial and the current would still behave as though they were delivering 100°C.

In this system even if the temperature in the furnace is unsatisfactory (incorrect), it can in no way alter the input to the furnace control to compensate. Or, put into control terms, *the output quantity has no effect on the input quantity*. In such a case the system is identified as an **open loop control system**. The adoption of open loop control requires very careful consideration since, as illustrated, any change in external conditions may cause the output of the system to fluctuate, or drift, in a manner that cannot be tolerated by the application concerned. It would be intolerable, for example, for a machine tool slide to indicate a movement of 100 mm and only actually move say 99 mm. This could happen using open loop control.

3.1/1 Block diagrams

By convention, control systems are represented on paper by **block diagrams**. This allows any system, regardless of power requirement, to be visualised simply and clearly. It is often known as the **black box** approach since a detailed knowledge of the workings of the component parts of the system is not required. It is only necessary to know how the output signal will respond to a given input signal and not what actually happens inside the box.

Fig. 3/2 Block diagram of open loop control of furnace temperature

POWER SOURCE | TEMP. DIAL SETTING | HEATING ELEMENT | FURNACE TEMP.

The block diagram of the open loop temperature control system, described above, is shown in Fig. 3/2.

3.1/2 Feedback

A thermometer (pyrometer) can be added to the system for the purpose of indicating the value of the temperature of the furnace. A human being can then read the thermometer and adjust the temperature dial to increase or decrease the indicated furnace temperature as desired. We have now introduced *feedback* into the control system by specifying and building in a **feedback loop**.

The *output quantity* (temperature) *is now having an effect on the input quantity* (although only manually). This system is now classified as a **closed loop control system**. More precisely, the operator reads the thermometer, compares the reading with the requested value, and compensates accordingly.

This is still manual control. To supply automatic control the thermometer can be replaced with a thermocouple. A thermocouple is a device that produces a voltage which is proportional to temperature. This voltage is an ideal form of feedback since it can easily be sensed and measured by simple instruments. Since we already have a reference voltage (the original input signal) it should be possible to compare them electrically. If there is a difference between them then we can assume that the actual temperature is different from the commanded temperature.

The actual control of the heating element thus depends on the *error* or difference between the desired and the actual temperatures. The system is said to be **error-actuated** and, since the actual value is subtracted ($-$) from the desired value (to determine the error), it is said to employ **negative feedback**. That is, we have an automatic closed loop control system employing negative feedback.

Fig. 3/3 Block diagram of automatic closed loop control of furnace temperature

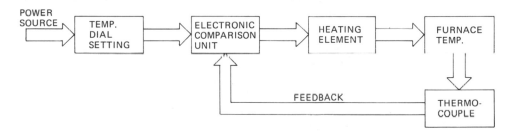

A block diagram of this automatic closed loop control system is shown in Fig. 3/3.

3.1/3 Closed loop CNC control

If this principle is now applied to CNC control, a program instruction becomes the command signal, an axis motor the controlled device, and the slide or axis position the controlled quantity.

In physical terms, the command signal itself is unlikely to be large enough to "drive" an axis motor and would, in practice, have to be magnified by some sort of amplifier. The amount of this magnification is termed **gain** or, more correctly, *loop gain* and becomes very important in control system design. There will, in the case of a closed loop system, also need to be some means of monitoring slide position and some means of comparing input and

Fig. 3/4a Block diagram of open loop control of machine tool axis

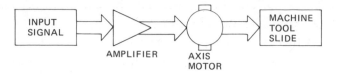

Fig. 3/4b Block diagram of closed loop control of machine tool axis

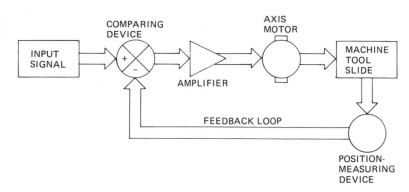

output quantities, i.e. providing feedback. Fig. 3/4 shows simplified block diagrams of typical open and closed loop control systems as applied to a controlled axis on a CNC machine tool.

Already we have built up a simplified model of a closed loop control system for a single axis of a CNC machine tool, and yet we do not require any detailed knowledge of the individual component parts that actually make the axis function. Such is the value of employing block diagrams in control system design.

The negative-feedback (error-actuated) closed loop concept has become the foundation for automatic control system design and is now widely applied in CNC control situations. The term **servo-control** is often used to describe such a system when applied to machine tool axis control. In fact, any fully automatic closed loop system, utilising some form of magnification in which mechanical position is the controlled quantity, is known, more precisely, as a **servomechanism**.

We shall see too, however, that open loop control systems also have their part to play in CNC control applications.

Closed loop systems, of necessity, require more component parts, and extra control circuitry, in order for them to perform the feedback function, and they are consequently more complex than their open loop counterparts. This inevitably means higher costs for the design and implementation of closed loop systems on CNC machine tools. We would expect also that the performance (quality) of the machine tool would be correspondingly better to justify this increase in capital investment.

3.2 Fundamental problems of control

There are a number of fundamental problems associated with designing control systems that are fully automatic. The way in which these problems are overcome will determine, to a great extent, the final performance or "quality" of the machine tool. This section examines some of the considerations associated with these problems.

3.2/0 Accuracy

It is appreciated in engineering that nothing can be perfectly accurate. This is demonstrated by the fact that most engineering components are produced with reference to both **dimensional and geometrical tolerances** (BS4500 and BS308 Part III respectively). These are really statements of how inaccurately we are allowed to work. It is therefore more correct to say that something is accurate to within certain limits, where the limits are specified. A CNC control system is no exception. It shares these same limitations and is only accurate to within certain limits.

True accuracy will only ever be achieved by monitoring and controlling the position of the actual cutting tool edge relative to the established datums. Since this is impractical under present-day circumstances, discrete (and often ingenious) **measuring devices** have to be employed on machine tool elements that are relatively accessible.

The **accuracy** of the servo-control system therefore depends partly on the accuracy of the measuring device used to monitor the position of the machine slide, partly on how this measuring device is utilised, and partly on any mechanical inaccuracies present within the total system.

For example, it is more accurate to measure the linear motion of the machine tool slide than to measure the rotation of the leadscrews that drive the slide. This is because of losses (such as *backlash* and *torsional wind-up*) that contribute to positional inaccuracies. In both cases the measuring device may be totally accurate by manufacture, but the system accuracy is determined by how they are utilised and other mechanical imperfections.

A typical accuracy specification for a CNC milling machine, for example, would be $\pm 0.025\,\text{mm}/300\,\text{mm}$.

3.2/1 Resolution

The term **resolution** refers to the smallest increment, or dimension, that the control system can recognise and act upon. This is not the same as accuracy. For example, a measuring scale may have 20 divisions engraved upon it in which case its resolution would be 1 in 20. It may not be accurate, however, since the divisions could be unevenly spaced throughout its *range*.

When the range of a measuring device is large, overall resolution is sometimes obtained by utilising a **fine measuring device** (over a small range) backed up by a **coarse measuring device** to permit the full range of slide motion to be measured. This principle is firmly established in the design of the common workshop micrometer. The fine resolution device is the micrometer screw which measures over a range of 25 mm to an accuracy of 0.01 mm. If a greater dimension is to be measured then a coarse device, the frame, is also employed. The frames of larger micrometers go up in steps of 25 mm. Thus, the resolution of 0.01 mm can be retained over a full range of (large) measurements.

A typical resolution specification for a CNC milling machine would be 0.001 mm.

3.2/2 Repeatability

As already stated, perfect accuracy is unattainable and so some dimensional tolerance must be applied. The component will be considered "correct" if its dimensions lie anywhere within this **tolerance band**. If a certain slide position is commanded many times in succession (as with many parts being made from the same part program), there will be a difference, or scatter, in the positions actually taken up by the slides. This scatter is a measure of the **repeatability** of the system. The repeatability of a system will always be better than its accuracy.

Indeed, it can be a considerable advantage if the same, rather than varying, machining errors appear on each repetition part. For example, the expensive processes of fitting and inspection could largely be eliminated.

A typical repeatability specification for a CNC milling machine would be ±0.005 mm.

3.2/3 Instability

Closed loop control tends to make for more accurate performance since the negative feedback is continually trying to reduce any error to an acceptable level. Under certain conditions, however, this continuous corrective action can lead to instability. **Instability** is the tendency for the system to oscillate about a desired position. An axis slide travelling at high speed (say on rapid traverse) may have too much inertia to stop immediately it reaches the commanded position, i.e. when the error becomes zero. As a result it will **overshoot** its target and feed back an error signal. This error signal will cause a reversal of the slide motor to enable the slide to try and again reach the commanded position. If the slide then **undershoots** its target, the procedure will be repeated and so on. This condition is often referred to as **hunting**. It should be clear that instability is a function of closed loop control systems only, since it is entirely due to the characteristics of the feedback loop, the loop gain and the response of the system.

3.2/4 Response

The **response** of a control system is the time lag between the application of the input signal and the controlled condition reaching the desired value. The tuning of the response speed for any particular control system is a compromise between a minimal time lag and maintaining the stability of the system. It is a complex problem since to reach a commanded position from rest a slide first has to accelerate, achieve a **steady state** condition, and then decelerate onto the target with the minimum of overshoot. This has to be maintained under widely varying load conditions, i.e. under rapid traverse or feed, with an empty or fully loaded table.

3.2/5 Damping

To counter the effects of excessive overshoot or undershoot, and hence help minimise hunting, **damping** may be introduced into the system. A certain amount of damping will already be present within the system due to the effects of friction. Any undesirable oscillations will die out more quickly as the amount of damping increases. However, too much damping will cause the response of the system to be unnecessarily slow.

The effectiveness of damping is seen in things like car shock absorbers and swing door closers. More importantly we are probably also aware of the difference in behaviour of such systems should this characteristic of damping be removed!

3.2/6 Control engineering

Control system design is not easy. It relies heavily on the strict applications of complex mathematics and a sound knowledge of the laws of physics. In addition, expertise in mechanics, electronics, pneumatics and hydraulics must also be integrated to produce a working control system.

In fact, since around the 1940s control engineering has matured into being an engineering discipline in its own right. The treatment thus far has been strictly non-mathematical and has served merely to introduce some of the language, terms and concepts used in modern control system design. It is outside the scope of this book to take the analysis further.

3.3 Types of positional control

One way of classifying CNC machine tools, other than by machine type, is by the positional control applied to their slide movements. This approach defines three broad headings which are sufficient to classify all CNC related equipment.

3.3/0 Positional or point-to-point control

Positional control is employed where the machine tool slides are required to reach a particular fixed coordinate point in the shortest possible time.

Note that *no* machining takes place until all slide movement has ceased. Indeed, it is common for the slide/table to be firmly clamped prior to the actual machining operation. This means that very low friction bearing systems can be specified to move the table (i.e. air bearings) and that damping within the slideways is unnecessary. Thus, the work that point-to-point machines perform is at the specified location and *not* along the movement from one programming point to the next.

The path of getting from one point to another is not fixed or even important. Fig. 3/5 shows a typical tool path between programmed points on a point-to-point machine. At first it appears that the selected path between points is unpredictable and somewhat random.

Fig. 3/5 Point-to-point control follows a somewhat irregular straight line path

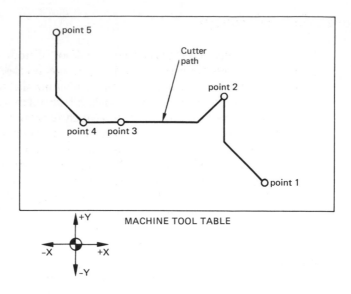

Fig. 3/6 Various ways of moving between two points

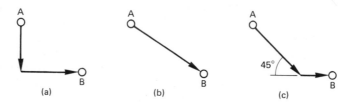

To understand the operation consider the various ways of moving between two points A and B. Fig. 3/6 shows three possible methods.

Method (a) is perhaps the slowest method in that each axis is energised separately. It has the advantage that a very simple control system would suffice since a coordinated movement of the axes is not required.

Method (b) is undoubtedly the quickest route between the points since it is working along the shortest path. This method, however, implies the use of sophisticated equipment to coordinate the speed of each axis to maintain a direct line. Such expense would probably not be justified on a simpler point to point machine.

Method (c) is the most common system. Here, both axes start to move simultaneously at full speed (hence a 45° path), and then to final position in one axis. Control requirements are kept simple without sacrificing too much speed. This path movement, however, could cause collisions with clamps or other projections that might fall within its programmable area. Reference back to Fig. 3/5 should now confirm the form of the path traced by the system.

Point-to-point control would be specified on equipment such as drilling and boring machine tools, punch presses, coordinate assembly machines (for PCB work), etc.

3.3/1 Paraxial or straight line control

Operations such as milling and turning, however, do dictate that the movement path between programmed coordinate points is important and, unlike drilling and boring operations, should be under full control at all times. Furthermore, speed control of the individual axes (commonly termed *feed*) must now also become a controlled variable in order for diagonal lines to be traced.

Note that this is the first time we have mentioned the problem of **feed control**. The block diagrams of the simplified models for open and closed control (introduced in section 3.1/3) will, in practice, need to be modified to incorporate arrangements for feed control.

There are probably no dedicated CNC **straight line control** metal cutting machine tools in widespread use. Straight line control is really a very limited form of contour control and to dedicate versatile computer control to such a menial task would be extremely wasteful. This system of control was more attractive on earlier NC machine tools since it was more versatile than point-to-point systems yet did not incur the increased costs of the more expensive continuous path systems. These controls utilised a method of moving from one point to another known as Linear Interpolation. Since all contouring controls provide linear interpolation, this will be more appropriately discussed in the following section.

3.3/2 Continuous path or contouring control

The method by which contouring systems move from one programmed point to another is called **interpolation**. This is a feature that merges the individual axis commands into a pre-defined tool path. There are three types of interpolation: *linear*, *circular* and *parabolic*. As stated previously most CNC controls now provide both linear and circular interpolation. Few controls used parabolic interpolation.

1 Linear Interpolation This means moving from one point to another in a straight line. With this method of programming any straight line path can be traced. This will include all taper cuts. When programming linear moves, the coordinates must be given for the end of each line only, since the end of one line is the beginning of the next. The *interpolator* within the control unit calculates intermediate points and ensures that a direct path is traced by controlling and coordinating the speeds of the axis motors. Circles or arcs can only be programmed with some difficulty. The circle or arc must be broken down into a number of straight line moves. The smaller each of these segments becomes, the smoother the circle or arc will be. This is illustrated in Fig. 3/7. The linear interpolation control requires that the end coordinate of each

Fig. 3/7 Circular arcs are made up of a series of straight lines

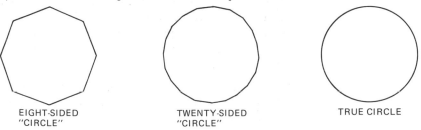

EIGHT-SIDED
"CIRCLE"

TWENTY-SIDED
"CIRCLE"

TRUE CIRCLE

of these segments be provided. It can be seen that trying to programme such curves with only linear interpolation results in specifying hundreds of coordinate positions and requires programs of vast length.

2 Circular Interpolation   The programming of circles and arcs has been greatly simplified by the development of circular interpolation.

Arcs up to 90° may be generated, and these may be further strung together to form half, three-quarter or full circles as desired.

Circular interpolation normally works from the current programmed position. The end point (X and Y coordinates) of the arc and the arc radius must also be specified. The circular interpolation control breaks the arc into small linear moves of high resolution.

Circular interpolation is limited to one plane at a time. Thus, a circle can be traced in the X-Y plane, the X-Z plane or the Y-Z plane but not in a combination of planes. Three-dimensional circular contouring is thus not possible using only circular interpolation. It may, however, be required for some applications and must be accomplished in other ways. See Fig. 3/8.

It is nonetheless a very versatile function and many free form shapes can be closely approximated to a series of arcs.

Fig. 3/8a Circular interpolation is limited to one selected plane at a time

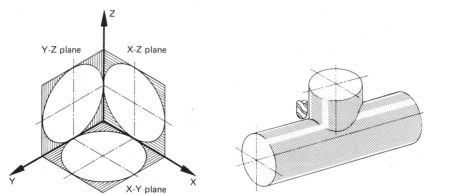

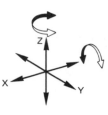

Fig. 3/8b Three-dimensional circular contouring

3 Parabolic Interpolation   The third method of interpolation, parabolic, is especially suited to mould and die manufacture where free form designs, rather than strictly defined shapes, are the norm. It easily adapts to those applications where aesthetic appeal is more important than mathematical description.

Parabolic interpolation positions the machine between three non-straight line positions in a movement that is either a part or a complete parabola. Its advantage lies in the way it can closely approximate curved sections with as much as 50:1 fewer points than with linear interpolation. A parabola is shown in Fig. 3/9c.

Such interpolation methods are ideally suited to the fast and accurate calculation abilities of the present-day digital computers. Most of the versatility of CNC machine tools is due to the undoubted flexibility resulting from such techniques. Certainly, programming components without them would be a long and arduous task.

All interpolation techniques find their origins in fundamental mathematical analyses of the functions on which they are based. The basis of each interpolation method is illustrated graphically in Fig. 3/9.

Fig. 3/9 Fundamental
basis of linear, circular
and parabolic
interpolation

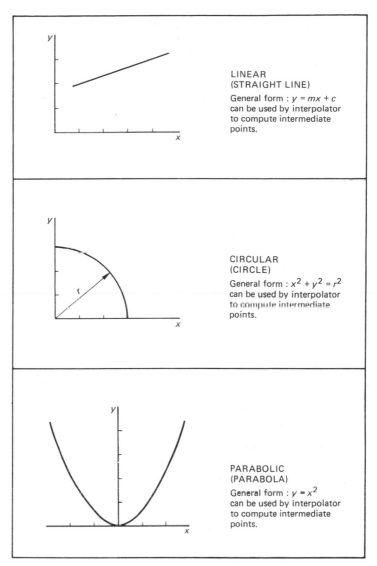

LINEAR
(STRAIGHT LINE)

General form : $y = mx + c$
can be used by interpolator
to compute intermediate
points.

CIRCULAR
(CIRCLE)

General form : $x^2 + y^2 = r^2$
can be used by interpolator
to compute intermediate
points.

PARABOLIC
(PARABOLA)

General form : $y = x^2$
can be used by interpolator
to compute intermediate
points.

3.3/3 Control axis classification

BS3635 Part 1:1972 suggests a means of classifying the capabilities of numerically controlled machine tools based on control capability.

Positional, straight Line or Contouring systems are identified by the letters
P, **L** and **C** respectively. These identifying letters are preceded by the number
of axes under such control. For example:

2P,L   indicates a machine where two axes have positional control and one
axis has linear control. This would suggest a point-to-point drilling machine.

2C,L   indicates a milling machine having continuous path control (in X
and Y), with linear positional control (of feed and depth), in a third (Z)
axis.

You may find it interesting to speculate how a CNC turning centre may be
classified under this system, or what machine tool is indicated by a classification
4C,L?

3.4 Machine tool control

3.4/0 Practical aspects of control

The practical aspects of control applied to CNC machine tools fall broadly into three categories:

a) Control of spindle speed.
b) Control of slide movement and velocity.
c) Control of slide position.

The following discussion introduces those factors that influence how these aspects are dealt with on CNC machine tools.

3.4/1 Control of spindle speed

The desirable attributes of spindle drives include the following.

a) They should be capable of supplying a wide range of spindle speeds to accommodate differing workpiece sizes and materials.
b) The speed range should be infinitely variable between the upper and lower limit to maintain constant cutting speeds.
c) High-power output should be available over the entire speed range to deliver constant torque, especially under low-speed/high-load cutting conditions.
d) They should be small and compact to be integrated into the machine tool structure, and exhibit quiet vibration-free operation.

Spindle drives are almost universally provided by either DC (direct current) or AC (alternating current) electric motors. The majority of CNC machine tools utilise **DC motors**.

Drive arrangements may be **direct**, from the spindle of the motor, or **indirect** via belt or geared transmissions.

Belt drives will be quieter and transmit less vibration. If geared drives are employed, it is common for the gears to be in constant mesh and selected remotely by electro-magnetic or hydraulic clutches. In this way programmable speed changes may be accomplished from within the part program. Such systems are less common since relatively high forces are required to effect gear changes and inertial considerations limit the rate at which such changes can be accomplished.

If direct drives are utilised, stepless speed variation can be accomplished entirely by electrical means. In the case of DC motors, speed is varied by altering the applied voltage; in the case of AC motors, the speed is altered by varying the frequency of the supply. Direct drives have the added advantage of eliminating the need for clutches and gearboxes to provide a simple, reliable and responsive performance requiring minimal maintenance.

It is desirable that there should be no manual intervention for the selection or changing of speed range since unmanned operation is one goal of CNC.

3.4/2 Control of slide movement and velocity

To accurately control machining conditions and cutter path, both positional and velocity information is required.

Control of both slide movement and velocity is determined by the type of control system employed. Consideration must be given to both open loop and closed loop systems.

OPEN LOOP CONTROL SYSTEMS By definition, **open loop control systems** do not employ feedback. This implies that neither the movement nor the velocity of the slide is being measured.

To accomplish accurate control of both movement and velocity in open loop control systems a special motor known as a **stepper motor** is employed. The stepper motor is unique in that it does not need the application of a varying (analog) voltage, as do conventional AC or DC motors. It is a digital device.

The principle of the stepper motor is that, upon receipt of a digital signal (a *pulse*), the spindle will rotate through a specified angle (the *step*). The step size is determined by the design of the motor but will typically be between 1.8 and 7.5 degrees. Thus, if two digital pulses are applied, then the spindle rotor will rotate by 2 steps, or by between 3.6 and 15 degrees depending on the motor design. Thus, by counting (electronically) the number of pulses sent to the stepper motor, and by knowing the lead of the axis leadscrew, the distance traversed can be accurately predicted. There is no need for positional feedback.

Velocity of the axis movement is determined by how quickly the pulses are sent to the stepper motor (the *pulse frequency*). If the pulses are sent very rapidly, then the feed-rate will be high; if the pulses are sent very slowly then the feed-rate will be low. The speed at which the pulses are transmitted can be accurately governed by the CNC control system. There is thus no need for velocity feedback.

In practice, digital switching circuitry and some means of power amplification are required to drive the stepper motor. This collection of electronics is known as a *translator*.

Advantages of employing stepper motors may be summarised thus:

a) The total drive system is considerably simplified since positional and velocity feedback do not have to be provided.
b) The cost of the drive system is considerably reduced.
c) Because of the absence of feedback the instability problem of hunting is eliminated.
d) Maximum torque is available at low pulse rates so acceleration of loads is accomplished easily.
e) When command pulses cease, the spindle rotor stops and there is no need for clutches or brakes.
f) Because the motor remains energised when in a stationary position it will inherently resist dynamic movement, up to the limit of its holding torque. For this reason stepper motors tend to run hot even when stationary.
g) When power is removed, the motor is magnetically detented in its last position.

h) Multiple stepper motors driven from the same pulse source maintain perfect synchronisation.

i) Control is very easy since it is a digital device and compatible with the output from the CNC control unit.

Stepper motors do however have certain limitations that restrict their usage in CNC machine tool applications. These limitations may be summarised thus:

a) Power output of stepper motors is relatively low, and this restricts their application to the control of smaller, lighter-duty machine tools.

b) Pulse rates are restricted to about 10 000 pulses/sec (10 kHz) which restricts maximum speed of axis movement to about 1/5 that attainable by closed loop systems.

c) If the axis movement is stalled, the pulses continue to "count" and loss of position will occur.

CLOSED LOOP CONTROL SYSTEMS **Closed loop control systems**, by definition, require both positional and velocity feedback. Since the control system is not required to count pulses there is a greater choice of spindle drives. AC or DC **servomotors** are normally employed.

Ideal closed-loop axis-drive servomotors should exhibit the following characteristics:

a) Reversal of direction of rotation when required.

b) Torque output should be proportional to speed.

c) High initial starting torque.

d) Fast and accurate starts, stops and spindle reversals.

e) Mechanically compatible with the machine and electrically compatible with the controller.

DC servomotors offer higher power output than AC servomotors for a given physical size. They combine high torque capability, high acceleration and low inertia for optimum system response. Speeds in excess of 3000 rev/min and as low as a few rev/min can be achieved without stalling. DC servomotors are less expensive than AC servomotors although they require more maintenance (notably to brushes and commutators). DC servomotors are more likely to provide radio interference during operation. The fact that both types of motor are electro-magnetic is used to provide regenerative braking to cut down deceleration times and minimise axis overrun.

Both types of servomotor exhibit an increase in speed with an increase in input signal. However for any given increase in the input signal, the speed of the motor cannot be accurately predicted. Extra arrangements have to be made to measure the speed of the motor and to compare this actual speed with the commanded speed. In control terms, **velocity feedback** must be provided. Velocity feedback is provided by building in, normally within the servomotor case, a device called a **tachogenerator**. A tachogenerator is a device that gives out a voltage that is proportional to its speed. This voltage is used as the feedback to monitor motor speed and hence axis speed or feed.

3.4/3　Control of slide position

The ideal means of measuring slide position would be to continuously measure the position of the cutting tool edge relative to the program or machine tool datum. Under such conditions no losses would be incurred and tool wear would automatically be compensated for. Unfortunately, this has not yet been achieved. Measurement from the cutting tool edge is made almost impossible by the presence of swarf, coolant, holding devices and, in many cases, the component itself. **Positional feedback** is provided by measuring slide movement indirectly via compact and discrete measuring **transducers** attached to both fixed and moving machine tool members.

A transducer is a device that converts energy in one form into energy in another form. For example, a thermocouple converts heat energy into electrical energy, to give a voltage that is proportional to temperature. A tachogenerator is a transducer that converts angular velocity into voltage.

Basically, two types of position-measuring transducer are employed on CNC machine tools: ANGULAR Transducers and LINEAR Transducers. Such transducers must be capable of measuring position (from a prescribed datum) or distance moved (incremental or absolute) from some reference point. Each controlled axis requires a position measuring transducer.

Position-measuring transducers may be either *analog* or *digital* in their design. An analog quantity is one that can vary continuously, between limits, over a period of time. Voltage, temperature, sound, etc. are analog quantities. By contrast, digital devices operate at two (and only two) distinct states. These two states may be referred to in a variety of ways: high/low, on/off, 1/0, set/reset, mark/space. There can be no in-between state. In digital systems the "high" state is represented by the presence of a voltage and the "low" state by the absence of a voltage. By convention these states are most commonly represented by 1 and 0 respectively.

3.4/4　Angular position measuring transducers

Angular transducers operate by measuring angular rotation, normally of the axis leadscrew. By knowing the lead of the leadscrew, the movement of the worktable can be ascertained. The most popular angular transducers are discussed below.

SINGLE RADIAL GRATING　In this system a transluscent disc is ruled with a series of **radial** lines (**gratings**). The resulting disc is made up of alternate (uniform) transparent and opaque areas. This disc is then keyed to the axis leadscrew. A collimated (parallel) light source and photocell arrangement is mounted such that, with rotation of the leadscrew, the photocell will sense alternate light and dark areas. The set-up is shown in Fig. 3/10a.

As a dark area of the disc is gradually uncovered, the light intensity falling on the photocell gradually builds up until it reaches a maximum when it is completely uncovered. As the disc continues to rotate, the following dark area starts to impede the light intensity falling on the photocell. The light intensity will gradually reduce until it is zero, when the dark area again cuts out light transmission. Since the photocell gives out a voltage that is proportional to the intensity of the light it receives, the resulting output will resemble

Fig. 3/10a Single radial grating position transducer

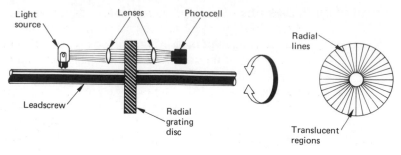

Fig. 3/10b Output from photocell transducer depends on light intensity

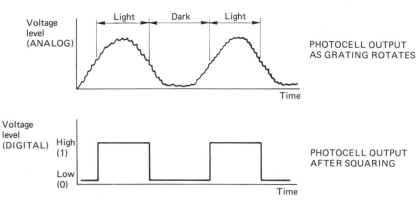

Fig. 3/10c Two photocells sense direction of movement by phase difference

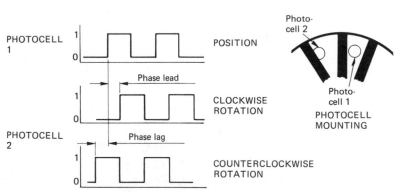

the shape of a sine wave. An electrical component known as a Schmitt Trigger converts this sinusoidal-shaped output into more of a square-shaped (or pulsed) output. This is illustrated in Fig. 3/10b.

The output can now be recognised as a series of discrete pulses, each pulse corresponding to a transparent region of the disc. Each pulse represents an angular movement of the leadscrew. By knowing the number of lines (hence the number of transparent areas) engraved on the disc, and the lead of the axis leadscrew, the movement of the worktable can be calculated by counting the number of pulses sensed.

A second photocell must be utilised to sense the direction of rotation. By positioning the second photocell as shown in Fig. 3/10c, the pulsed output will be identical to that of the first photocell. It will however be slightly out of step. The degree of this out-of-step is termed the **phase difference**. The phase difference can be sensed to determine the rotation of the leadscrew. This is also illustrated in Fig. 3/10c.

Accuracy of position measurement by this method is limited by two main factors. Firstly, there is a physical limit to the number of lines that can be engraved on a given disc, and secondly, the photocell itself requires a gap of a certain size for it to sense variations in light intensity.

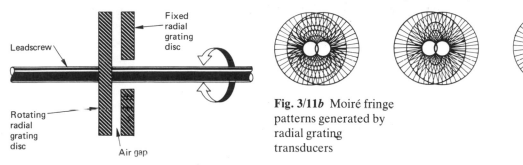

Fig. 3/11a Radial Moiré fringe grating position transducer

Fig. 3/11b Moiré fringe patterns generated by radial grating transducers

RADIAL MOIRÉ FRINGE GRATINGS Two radial gratings can be positioned adjacent to each other. One grating is fixed and the second is keyed to (and rotates with) the leadscrew. The arrangement is shown in Fig. 3/11a. If the disc centres are offset, a moving pattern of lines is produced when the shaft rotates. This pattern is referred to as a **Moiré Fringe** and is illustrated in Fig. 3/11b. Observation shows that if there are n-lines engraved on the disc, this pattern will rotate n-times during each revolution of the shaft. This is a form of optical magnification which can be sensed by the use of photocells as in the case of a single grating. Direction of rotation will be detected by using a second photocell.

Fig. 3/12 Binary shaft encoder for positional feedback

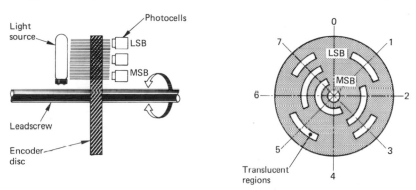

SHAFT ENCODER In this arrangement a disc is encoded by etching transparent and opaque (light and dark) regions in the form of a binary pattern. (Consult Chapter 4 for a detailed treatment of the binary system.) Essentially at any angular position of the encoder disc a series of photocells will transmit a binary code depending on the form of the light and dark areas sensed. The binary code transmitted is an indication of the absolute position of the leadscrew throughout its rotation. A **shaft encoder** is illustrated in Fig. 3/12. The accuracy of the encoder disc is a function of how many photocells (binary digits) are used to make up the binary code. Although three are shown in Fig. 3/12, up to twelve may be employed.

The binary code at each of the eight radial positions (corresponding to the decimal numbers represented at those positions) shown in Fig. 3/12 is generated by a unique combination of photocell ONs and OFFs. The resulting code is shown in the table below (1 represents ON and 0 represents OFF by convention). In practice, malfunctions can occur if the photocells become skewed from the radial line. At the transition points, between (decimal) numbers 1 and 2, 3 and 4, and 5 and 6, a number of photocells change their states at once. This can be another source of malfunction. This condition becomes more common as the number of binary digits (photocells) increase. To overcome this problem the natural binary code is modified so that, at any transition point, a change in only one binary digit is required. The resulting code is known as the **Gray code**. Gray coded discs increase reliability but require the use of additional decode circuitry within the transducer.

An extra photocell used in conjunction with a single notch in the radial discs will also be provided to "count" the number of complete revolutions of the disc/leadscrew.

Decimal Number	Natural Binary Code	Gray Code
0	000	000
1	001	001
2	010	011
3	011	010
4	100	110
5	101	111
6	110	101
7	111	100

SYNCHRO and SYNCHRO RESOLVER   The **synchro transducer** utilises the principle of magnetic induction. The principle of magnetic induction is as follows. If a conducter carries an electrical current, a magnetic field is produced around that conductor. The situation also exists in reverse in that, if a conductor is moved in the vicinity of a magnetic field (or vice versa), a current will be induced in that conductor.

A series of stationary windings are arranged around the periphery of the synchro. This is termed the *stator*. A central spindle, also carrying a winding, is attached to the leadscrew and rotates within the stator. This is termed the *rotor*. The stator windings are supplied with a constant voltage which sets up a magnetic field around them. As the rotor rotates, with the leadscrew, a voltage is induced in the rotor winding. This induced voltage varies from a minimum to a maximum, depending on its position relative to the stator windings, in a sinusoidal fashion. The magnitude (size) of this voltage represents the angular position of the rotor. By counting the number of complete revolutions of the leadscrew, sensing the angular position of the rotor, and relating this to the lead of the axis leadscrew, the movement of the worktable can be established.

A **synchro resolver** works on a similar principle but has two stator windings positioned at right angles to each other. The windings are fed with an identical AC voltage but shifted in phase (out of step) by 90°. The induced voltage in the rotor winding will itself have a phase shift, with respect to the stator windings, which is determined by its angular position. In a synchro resolver it is the phase shift of the induced voltage that is proportional to the angular position of the rotor.

3.4/5   Linear position measuring transducers

Position measurement by angular transducers is indirect in that the output has to be converted to actual table movement. This represents an additional source of loss or error. Linear measuring transducers operate by recording actual movement of the machine worktable. Errors due to backlash and torsional wind-up, and leadscrew pitch errors, are eliminated. They have the disadvantage that they must be physically protected.

Fig. 3/13 Linear grating position-measuring transducer

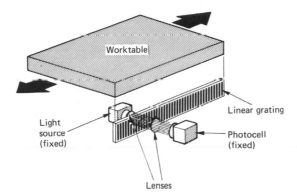

LINEAR GRATING   A **linear grating** works on the same principle as the single radial grating. A precision linear scale engraved with close-spaced parallel lines is fixed to the moving worktable. A light source and photocell detector are fixed on a convenient stationary element of the machine tool. This is illustrated in Fig. 3/13. As the transparent regions of the linear grating expose the light source, a pulse is registered by the photocell. By knowing the pitch of the engraved lines on the linear grating, and counting the number of pulses, the movement of the worktable can be established.

As with radial gratings, linear gratings utilise a second photocell to detect direction of movement. Linear gratings can be either etched onto glass, in which case light is detected through them, or engraved onto stainless steel tapes, in which case light is reflected from them.

LINEAR MOIRÉ FRINGE GRATINGS   Linear gratings may be limited in resolution for some applications. This is because of the physical limitation of the number of parallel lines that can be engraved, per unit length, to preserve a sufficient gap as required by the photocell. Moiré fringe gratings, working on the same principle as the radial Moiré fringe grating, can be arranged in a linear fashion.

A fixed-scale grating is attached to the machine tool structure. A smaller index grating is affixed to, and moves with, the worktable. The lines on the index grating are at the same pitch as those on the fixed grating, but are inclined at a slight angle. When the two gratings move relative to each other, the characteristic Moiré fringe pattern is observed to move across the grating. The intensity of the light falling on the photocell through the gratings will vary sinusoidally as movement proceeds. A number of photocells can be arranged to monitor the same fringe movement to increase resolution. The order in which the photocells are excited (monitored by phase difference) will determine the direction of motion of the worktable.

Moiré fringe gratings can easily be reproduced to examine the effect. Simply rule some closely-spaced parallel lines on a sheet of paper. Reproduce this on a sheet of transparent film and experiment by placing the film over the paper, at a slight angle, and moving it from right to left. The fringe can be observed to move across the grating. Notice that reversing the relative direction of movement causes the fringe to move in the opposite direction.

INDUCTOSYN   The **inductosyn** operates on the synchro-resolver principle but is effectively "laid out flat".

It comprises a long fixed scale carrying a pattern of wires that repeat at regular intervals (every 2 mm or 3 mm). This performs the same function as the rotor winding in the synchro resolver. The winding is usually printed onto a glass scale. Glass is an insulator, is very stable, and has a linear coefficient of expansion. A second, smaller scale, carrying two similar patterns of repeating windings, is mounted as a slider on the moving worktable. The slider performs the same function as the stator windings in the synchro resolver. The two windings of the slider are supplied with identical AC voltages separated in phase by 90°. The fixed-scale winding will have a voltage induced in it. The difference in phase between the induced voltage of the fixed scale and the supply voltage of the slider is a measure of the positional movement of the worktable.

A big advantage is that one cycle of output voltage can easily be divided electronically (up to 1/500), for greater resolution. For practical purposes, most inductosyn scales are produced in sections. A number of sections are then fitted and aligned to accommodate the full axis movement. Alignment of windings is not crucial since it is a magnetically coupled voltage that is being measured. The inductosyn is illustrated in Fig. 3/14. Since there is no contact between the elements there is no wear. Physical protection must, however, be provided.

Fig. 3/14 Components of the inductosyn linear position transducer

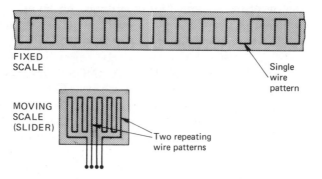

To effect feedback, the output signals from each analog-type positional transducer will need to be converted into a form the control unit can accept. Since CNC control units are digital devices this form must be digital. Additional electronic circuitry will thus be required to compare command signals and feedback signals. Such electronics will employ standard devices such as *analog-to-digital convertors* (ADC) or *phase-to-digital convertors* (PDC). The resulting output can then also be easily displayed in digital form for axis readouts.

All of the above transducers are indirect in that they do not indicate the precise position of the tool cutting edge in relation to the workpiece. Slight losses or errors are therefore somewhat inherent. Many efforts are made to reduce such losses to an absolute minimum. For example, inappropriate location of the measuring transducers will contribute to likely losses within the total control system. Linear transducers should be mounted near to sliding surfaces and leadscrews, observing the need for accessibility for maintenance purposes. Losses in angular measuring transducers will be minimised if they are mounted at the free end of the leadscrew rather than at the driving end.

Questions 3

1   What are block diagrams and why are they so useful?
2   Explain, using block diagrams, the terms "open loop" and "closed loop" in the context of CNC control systems.
3   Explain what is meant by the term "feedback". State *two* types of feedback that will be required in a closed loop CNC control system.
4   Define the term "hunting" and explain why it can occur.
5   Briefly explain the differences between "point to point", "straight line" and "contouring" control in the context of CNC machine tools. State a machine tool using each type of control system.
6   What is the difference between linear and circular interpolation, and when would each be employed?
7   Define the terms "repeatability", "accuracy", "resolution" and "damping" in the context of CNC control systems.
8   Control system capability on CNC machine tools can be designated by the letters P, L and C. Briefly explain what each letter stands for and state the type of machine that is indicated by the classification 4C,L.
9   Discuss the relative advantages and disadvantages of using stepper motors and DC servo motors for the control of slide movement on CNC machine tools.
10   What is a transducer. Illustrate your answer by quoting typical examples that would be used on a CNC machine tool.
11   Briefly explain the difference in operation of a "shaft encoder" and a "radial grating" transducer used for positional feedback.
12   State *three* different types of linear measuring transducer that can be used for positional feedback, and briefly explain their principles of operation.
13   Where would "backlash" and "torsional wind-up" occur on a CNC machine and how do they influence control system design?
14   Briefly outline the factors that should be considered in the choice of a suitable machine tool spindle drive arrangement.
15   What is a tachogenerator and where would it be used on a CNC machine tool?
16   Explain how direction of movement is sensed when using grating-type positional measuring transducers.
17   What is a Moiré Fringe?

18 Define the term "error-actuated negative feedback" in the context of closed loop control system operation.

19 Why are all current closed loop system designs somewhat of a compromise in terms of positioning accuracy?

20 CNC machine tools will use both digital and analog signals. Give an example of where each signal type would be employed.

Computer considerations   4

4.1   Computer systems

4.1/0   What is a computer?

Modern computers are devices configured to manipulate information (**data**) according to a set of rules in a very precise and ordered manner. The set of rules by which the computer operates is called a **program**. Much of the flexibility of computer systems is derived from the fact that different programs for handling the same data (from a common database) can be designed, and run on a single computer system, or that a single program may work on many different sets of data (a post-processor operating on a collection of different part programs). *All* the actions the computer takes are predetermined by the programs it runs; it cannot display "intelligence" of its own (yet).

The principal advantages of employing computer systems may be summarised thus:

They can work at incredible speeds.
They can work consistently without requiring rest, supervision or other services.
They are extremely accurate.
They do not forget.
They can issue warnings and reminders based on the most up-to-date information.
They are compact and relatively cheap.

These factors translated into industrial terms mean greater productivity, lower costs, consistency of output and tighter controls on the day to day running of the organisation.

4.1/1   Computer sub-systems

A computer system may be designed to be a general-purpose data-handling system. For example, a typical home or business computer may display immense versatility using colour and graphics. Alternatively, it may be designed to carry out a very specific task. Such a computer system is then known as a **dedicated computer**. Dedicated computers may be found in washing

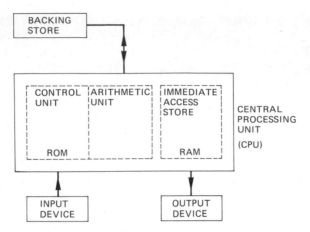

Fig. 4/1 Block diagram of a typical computer system

BACKING STORE

CONTROL UNIT   ARITHMETIC UNIT   IMMEDIATE ACCESS STORE

ROM   RAM

CENTRAL PROCESSING UNIT (CPU)

INPUT DEVICE

OUTPUT DEVICE

machines, cash registers, petrol pumps, etc., and in the machine control units of CNC machine tools.

No matter what their intended purpose all computer systems have a common structure. The concept of block diagrams was introduced in Chapter 3 (section 3.1/1). Using the same approach a block diagram of a typical computer system is illustrated in Fig. 4/1.

The system comprises a **central processing unit** or CPU which communicates with various **peripheral devices.** The CPU is itself made up of three distinct sub-sections:

a) The **control unit** is responsible for coordinating all the functions carried out by the computer. These functions include processing the current program instructions in the correct order at the correct time, and also looking after the various "housekeeping" tasks of loading in and saving programs, handling error conditions, etc.

b) The **arithmetic unit** is responsible for any calculations but can also select, sort and compare information which can be used to execute some parts of the program in preference to others.

c) The **immediate access store** is the **internal memory** of the computer used to store the program and the results of any calculations performed by the arithmetic unit. Its operation is electronic and, since there are no moving parts, data can be accessed in nanoseconds (one thousand millionths of a second).

In a CNC control system, the **input devices** will be many. Firstly, the part program must be entered into the control system memory. This may be done via the console keyboard, a punched tape reader, a magnetic tape reader, or a host computer. Secondly, the control system will also be monitoring the status of the machine tool itself in terms of spindle speed, feed rate, axis positions, etc., through various transducers. Similarly, the control system also has many **output devices** which it must address: the console display, punched tape punches, magnetic tape/disc units, as well as spindle and axis motors, coolant pumps, automatic tool changers, etc.

The wide variation in peripheral devices that need to communicate with the controller inevitably leads to some incompatibilities in the form in which each device accepts and transmits its information. To overcome this it becomes necessary to provide an **interface** between the control unit and the various peripheral devices. An interface is a collection of electronic circuitry designed to make information in one form compatible with information in another form.

4.1/2 What is a microprocessor?

A **microprocessor** is basically the "brain" of a digital computer system and is equivalent to the Control Unit and the Arithmetic Unit combined. It is a complex of micro circuits housed in a single integrated circuit or chip. A typical microprocessor can obey about 70 simple standard instructions and these are usually combined to form powerful programs. Since each instruction can be executed in a mere 2 to 10 microseconds (2–10 millionths of a second) complex tasks can be performed very rapidly.

An indication of the size and speed of the microprocessor is given in Fig. 4/2.

Fig. 4/2 Indication of the size and speed of a microprocessor

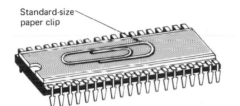

Standard-size paper clip

BY THE TIME YOU HAVE READ THIS A MICROPROCESSOR HAS CARRIED OUT 400 000 INSTRUCTIONS

The flexibility to configure and program a microprocessor to perform practically any function attracts system designers in diverse fields. The ability to change the function of a piece of equipment by changing some instructions in the microprocessor instead of redesigning a printed circuit board has obvious attractions. Thus, CNC control system suppliers are able to offer control system updates, at nominal cost, as extra facilities become available.

It is the use of the microprocessor that gives the microcomputer its name and makes the application of compact, low-cost computing power possible. However, a microprocessor is of little use on its own and must be combined with some form of input, some form of output, and, in most cases, extra memory, to form a usable computer system.

4.1/3 Computer memory

A computer is only capable of doing one thing at a time. The result of a calculation, for example, must be stored away in the computer's internal memory before the next program instruction is carried out, otherwise it will be lost.

Computer memory can be visualised as rows of storage locations or "pigeon holes". Each storage location is capable of storing a single **character** of information. Characters may be alphabetic letters, numbers or punctuation marks (termed *alphanumeric* characters).

The action of storing (or saving) a character in a memory location is termed **writing to** memory, and the action of fetching a character from a memory location is termed **reading from** memory. When a character is read from a memory location the contents of the location remain unchanged. Thus, data may be read any number of times. This is why a part program held in memory may be executed any number of times without reloading it. When a character is written into a memory location, it will overwrite the original contents of that memory location. This makes it possible to edit CNC part programs held in the memory of the control unit and thus make minor modifications to the program, at the machine.

The smallest unit of memory is one character or *one byte* of information. Memory capacity is usually quoted in terms of **kilobytes** or K for short. In the ISO number system we understand the prefix kilo- (or k) to represent multiples of 1000, e.g. 1 kilogramme (kg) = 1000 grammes, 1 kilometre (km) = 1000 metres, etc. In computer terms however kilo- represents multiples of 1024. The reason is as follows. The ISO system is a decimal-based system (it counts with a base of 10) and 1000 is a round multiple of 10, i.e. 10^3. Computer systems work in binary (which count with a base of 2) and successive multiplication by 2 cannot "hit" 1000 exactly. The nearest multiple is 1024 or 2^{10}. (See section 4.2 for a further explanation of the binary system.)

Computer memory devices fall broadly into two categories. Reference back to Fig. 4/1 shows that the control unit is also labelled ROM and the immediate access store labelled RAM:

ROM stands for **Read Only Memory**. As its name implies it can only be read by the processor. This type of memory has its contents built (or burnt) into it when it is manufactured. It is thus used to house the dedicated main control program used by the control unit. The contents of ROM cannot be erased even by removing the power supply. It is said to be a **non-volatile** memory device.

RAM stands for **Random Access Memory**. This is memory that can either be written to or read from at random. It is also termed read/write memory. The same memory locations can be used over and over again to hold different programs and data. Many CNC control systems allow more than one part program to reside in memory at the same time with the facility to switch between them. The contents of RAM are wiped out if the power supply is removed; it is thus termed a **volatile** memory device.

Typical CNC control systems have between 16 K and 64 K RAM.

Another popular, and very useful memory device is the **EPROM**. This stands for *Erasable Programmable Read Only Memory*. It is in fact a re-usable ROM. It can be programmed using a device called an eprom programmer. Once programmed the EPROM acts like a ROM in that its contents are non-volatile. An EPROM can be erased, if required, by exposing it to ultra-violet light for between 20 and 30 minutes. It can then be re-programmed. These devices make control system program updates extremely simple and cost effective.

Backing store refers to external storage such as magnetic tape or disc devices. Backing stores are necessary to provide permanent and backup copies of part programs for future use. Intelligent utilisation of backing stores can create easily accessible libraries of often used part programs, subroutines or macros. Backing store devices will be discussed more fully in section 4.4/2.

The general term **hardware** describes all the physical parts of a computer/CNC system. **Software** refers to computer or part programs and the media on which they are stored, and **firmware** describes computer programs that are stored on a chip, such as the dedicated main control program held, in ROM, within the CNC control unit.

4.2 The binary system and its importance

4.2/0 Number bases

In everyday arithmetic we make use of numbers expressed in the decimal (or denary) system. Decimal means "ten" and is called the base. A **number base** is simply the number of digits, including zero, which the system can use. Because zero must always be included, the actual digits used go up to one less than the base itself. Thus, in the decimal system the digits used are 0, 1, 2, 3, 4, 5, 6, 7, 8 and 9.

Very large or very small numbers are expressed using the same digits but in different positions. For example, the number 3256 means:

3-thousands + 2-hundreds + 5-tens + 6-ones (or units)

More precisely, each number is separated, in value, from its neighbour by a power of 10 (the base). The digit having the least value is at the right-hand end.

1000s	100s	10s	Units	position value
10^3	10^2	10^1	10^0	number base value
3	2	5	6	decimal value

Many other number systems are in use for specialist applications but they all have the common features outlined above, i.e. the use of a base and a value related to position which is a power of the base.

4.2/1 Binary numbers

A computer is an electronic device and as such relies on levels of voltage for its operation. If the computer were constructed to work according to the decimal system then it follows that it would require ten distinct voltage levels or states, one for each digit. In practice this is difficult to implement and it is far easier for electronic engineers to create electronic circuits using only two states. The two states then become: a voltage being present or no voltage being present; or an electronic switch being open or an electronic switch being closed. For this reason, the binary number system (with a base of 2) using only two digits, 0 and 1, is ideal for computer applications.

Because there can be no "in-between" state using this system it is referred to as being *digital*. The electronics employed is known as **digital electronics** and the computers that result are termed **digital computers**. CNC control systems employ digital computers.

Since only two digits are used, binary numbers take on the form of long strings of 1s and 0s, for example 101011. Each digit in the string is known as a **bit**. This comes from a contraction of the term BInary digiT. To understand the significance of such numbers we must revert to the theory of number bases introduced above. As with other number bases, the digit at the extreme right has the least value. The right-most digit in a binary number is thus known as the **least significant bit** (LSB). Conversely, the left-most digit, having the highest value, is termed the **most significant bit** (MSB).

Using the above analysis, the binary number 101011 can be translated as follows:

32	16	8	4	2	1	position value
2^5	2^4	2^3	2^2	2^1	2^0	number base value
1	0	1	0	1	1	binary value
MSB					LSB	

Note how the value of the bits increases by a power of 2 as we proceed leftwards. In fact the value of each bit is *double* that of its predecessor.

Since long strings of 1s and 0s can be somewhat confusing and difficult to evaluate, it is often more convenient for us to translate the binary number into its decimal number equivalent. From the above **bit pattern**, the binary number consists of:

$$(1 \times 32) + (1 \times 8) + (1 \times 2) + (1 \times 1) = 43$$

Thus, the decimal equivalent of 101011 is 43.

Exercise Calculate the decimal equivalent of the binary number 1101011 and the binary representation of 77.

4.2/2 Binary coded decimal

There appear to be two apparent drawbacks to the binary system as far as CNC work is concerned. The first is that as the value of the number increases so does the length of the bit pattern. The length of the bit pattern is known as the **word length**. Thus, large dimensions will require longer word lengths than small dimensions:

43 =	101011	6-bit word length
143 =	10001111	8-bit word length
943 =	1110101111	10-bit word length

Secondly, there seems no way of representing fractional numbers. In the decimal system we employ the decimal point. Digits to the left of the decimal points successively increase by a power of ten and numbers to the right of the decimal point successively decrease by powers of 10, e.g. 256.83.

Both these drawbacks are overcome, for CNC purposes, by using a system known as **binary coded decimal** (BCD). In this system each *digit* in the decimal number is given its own *binary word*. The separate words are then transmitted one after the other in the correct (decimal) order. For example, the number 256.83 expressed in BCD would read:

1st word	0010	2
2nd word	0101	5
3rd word	0110	6
		.
4th word	1000	8
5th word	0011	3

The decimal point is often not required for CNC applications, since many control systems supply the decimal point after reading the first 3 or 4 digits of the number. This can easily be sensed by the inbuilt computer logic of the controller, but it means that dimensions have to be "padded out" with leading zeros where necessary. For example, to input the dimension 2.56 mm into such a system would require the dimension to be expressed as 00256. Where control systems do not require this "padding", they are said to employ **leading zero suppression**.

It is a simple step to imagine the representation above being reproduced as rows of hole patterns in a punched tape. Where there is a binary 1, punch a hole; where there is a binary 0, do not punch a hole. Numerical information can thus be encoded onto perforated tape. The presence or absence of a hole can then be sensed by a variety of means and then decoded, according to the same principles, by computer logic within the CNC controller.

Binary coded decimal has added advantages especially when using punched tape as a program medium. Word lengths, and hence tape widths, can be kept to realistic sizes. Note how it is possible to represent all the digits, 0–9, using a word length of only four bits. This presentation also facilitates learning to recognise hole patterns for the digits 0–9. This enables operators to "read" the numbers, and hence the dimensions, punched in the tape.

4.3 CNC code systems

4.3/0 Alphanumeric characters

	bit 4	bit 3	bit 2	bit 1
1	0	0	0	1
2	0	0	1	0
3	0	0	1	1
4	0	1	0	0
5	0	1	0	1
6	0	1	1	0
7	0	1	1	1
8	1	0	0	0
9	1	0	0	1
10	1	0	1	0
11	1	0	1	1
12	1	1	0	0
13	1	1	0	1
14	1	1	1	0
15	1	1	1	1
16	0	0	0	0

Fig. 4/3 Sixteen possible codes using four bits

As stated earlier, a memory location is capable of holding a single character of information, and this character can be a digit, an alphabetic character or a punctuation symbol. In theory, we have just seen how numbers can be represented using binary coded decimal but what of the *alpha* characters? We shall see, in Chapter 7, that CNC control systems need alpha characters and punctuation symbols just as much as numbers.

If we examine the system of representing numbers in binary in the previous section a little more closely, it becomes apparent that each digit is merely a unique arrangement of 1s and 0s—a code! The numerical value assigned to any particular bit pattern just happens to coincide with the weighting associated with each position, starting from the right-hand side. Indeed, any arrangement of 1s and 0s can be chosen to represent any digit so long as the device **decoding** the number uses the same interpretation as the device **encoding** the number.

We used four bits to represent ten digits (0–9). In fact, the combinations of four bits will allow the representation of sixteen different symbols. Fig. 4/3 illustrates the possible combinations. It follows, therefore, that alpha characters and punctuation symbols could also be represented in coded form, by a unique bit pattern of 1s and 0s.

4.3/1 The ASCII code

The most widely used coding system for computer applications is the **ASCII** (pronounced ASKEY) code. This is an American-devised code and the letters stand for American Standard Code for Information Interchange. It is this coding system on which the ISO recommendations for CNC codes is based (BS3635 Part 1 :1972).

The ASCII code represents alphanumeric characters using a 7-bit word length. It is interesting to observe why a word length of 7-bits was chosen. For computing applications it will be necessary to represent, at least, the following characters:

26   upper case (capital) letters
26   lower case (small) letters
10   numeric digits (0–9)
 4   arithmetic symbols ($+, -, \times, \div$)
 1   decimal point
—
67   characters

The possible number of unique combinations that can be achieved using a 6-bit binary word length is 64 (or 2^6). Clearly, this is inadequate for even the minimum **character set** suggested above. A 7-bit word length is thus used which can accommodate many more characters: for example, a full range of punctuation marks and many commonly used symbols such as brackets, pound and dollar signs, percent and ampersand symbols, etc., with spare capacity for special **control codes** used for sending instructions to peripheral devices.

Exercise   Calculate how many characters, in all, can be represented using 7-bits.

Since computers are now commonly communicating with CNC machine tools directly, it demonstates sound judgement to base the CNC character codes on the established "standard" for the computer industry. (See also Direct Numerical Control in section 4.4/3.)

4.3/2   The ISO 7-bit numerical control code

BS3635 Part 1 :1972 specifies the standard **ISO 7-bit code** that is recommended for CNC applications by the International Standards Organisation. In fact, this code is a sub-set of the ASCII code, and comprises some 50 characters. It will be appreciated that the full ASCII character set would not be appropriate, in its entirety, for CNC. For example, there will be no need to specify lower case letters and many textual symbols for such a specialised application.

The full ISO character set, binary representation and decimal equivalents are shown in Fig. 4/4. Also shown is the ISO code set representation as it appears when punched in paper tape. Punched paper tape represents the single most popular storage medium for CNC machining applications. Indeed, so widespread was its use on the earlier NC machines that they were often referred to as tape-controlled machines. There are, however, numerous other means of data entry which will be discussed in section 4.4.

We shall use the representation in paper tape, of Fig. 4/4, to examine the make-up of the coding system. Remember: the presence of a hole represents a binary 1 and the absence of a hole represents a binary 0. We shall consider the vertical columns as *tracks* or *channels* on the tape.

Figure 4/4 warrants close scrutiny since, although the code primarily has a 7-bit word length, it is represented, on paper tape, by 8-bits. The most

DECIMAL EQUIVALENT	BINARY REPRESENTATION b7	b6	b5	b4	b3	b2	b1	NAME OF CHARACTER	CHARACTER SYMBOL	P	7 (b7)	6 (b6)	5 (b5)	4 (b4)	F	3 (b3)	2 (b2)	1 (b1)
0	0	0	0	0	0	0	0	NULL	NUL						•			
8	0	0	0	1	0	0	0	BACKSPACE	BS	•				•	•			
9	0	0	0	1	0	0	1	TABULATE	TAB					•	•			•
10	0	0	0	1	0	1	0	END OF BLOCK	LF					•	•		•	
13	0	0	0	1	1	0	1	CARRIAGE RETURN	CR	•				•	•	•		•
32	0	1	0	0	0	0	0	SPACE	SP	•		•			•			
37	0	1	0	0	1	0	1	PROGRAM START	%	•		•			•	•		•
40	0	1	0	1	0	0	0	CONTROL OUT	(			•		•	•			
41	0	1	0	1	0	0	1	CONTROL IN	)	•		•		•	•			•
43	0	1	0	1	0	1	1	PLUS SIGN	+			•		•	•		•	•
45	0	1	0	1	1	0	1	MINUS SIGN	−			•		•	•	•		•
47	0	1	0	1	1	1	1	OPTIONAL BLOCK SKIP	/	•		•		•	•	•	•	•
48	0	1	1	0	0	0	0		Ø			•	•		•			
49	0	1	1	0	0	0	1		1	•		•	•		•			•
50	0	1	1	0	0	1	0		2	•		•	•		•		•	
51	0	1	1	0	0	1	1		3			•	•		•		•	•
52	0	1	1	0	1	0	0		4	•		•	•		•	•		
53	0	1	1	0	1	0	1		5			•	•		•	•		•
54	0	1	1	0	1	1	0		6			•	•		•	•	•	
55	0	1	1	0	1	1	1		7	•		•	•		•	•	•	•
56	0	1	1	1	0	0	0		8	•		•	•	•	•			
57	0	1	1	1	0	0	1		9			•	•	•	•			•
58	0	1	1	1	0	1	0	ALIGNMENT FUNCTION	:			•	•	•	•		•	
65	1	0	0	0	0	0	1		A		•				•			•
66	1	0	0	0	0	1	0		B		•				•		•	
67	1	0	0	0	0	1	1		C	•	•				•		•	•
68	1	0	0	0	1	0	0		D		•				•	•		
69	1	0	0	0	1	0	1		E	•	•				•	•		•
70	1	0	0	0	1	1	0		F	•	•				•	•	•	
71	1	0	0	0	1	1	1		G		•				•	•	•	•
72	1	0	0	1	0	0	0		H		•			•	•			
73	1	0	0	1	0	0	1		I	•	•			•	•			•
74	1	0	0	1	0	1	0		J	•	•			•	•		•	
75	1	0	0	1	0	1	1		K		•			•	•		•	•
76	1	0	0	1	1	0	0		L	•	•			•	•	•		
77	1	0	0	1	1	0	1		M		•			•	•	•		•
78	1	0	0	1	1	1	0		N		•			•	•	•	•	
79	1	0	0	1	1	1	1		O	•	•			•	•	•	•	•
80	1	0	1	0	0	0	0		P		•		•		•			
81	1	0	1	0	0	0	1		Q	•	•		•		•			•
82	1	0	1	0	0	1	0		R	•	•		•		•		•	
83	1	0	1	0	0	1	1		S		•		•		•		•	•
84	1	0	1	0	1	0	0		T	•	•		•		•	•		
85	1	0	1	0	1	0	1		U		•		•		•	•		•
86	1	0	1	0	1	1	0		V		•		•		•	•	•	
87	1	0	1	0	1	1	1		W	•	•		•		•	•	•	•
88	1	0	1	1	0	0	0		X	•	•		•	•	•			
89	1	0	1	1	0	0	1		Y		•		•	•	•			•
90	1	0	1	1	0	1	0		Z		•		•	•	•		•	
127	1	1	1	1	1	1	1	DELETE	DEL	•	•	•	•	•	•	•	•	•

Fig. 4/4 Representation of the ISO numerical control code

significant bit (the leftmost track) is included, and reserved for, an error-checking device called a **parity check**. This is a system for detecting certain errors that could be present during data transmission and does not form part of the coding system itself. It is a system applied primarily to data stored, and transmitted, via tape media. The importance of the parity track is reflected in the widespread use of punched tape as the primary input medium. (Parity and error checking will be discussed in section 4.4/4.)

Let us first consider how the digits 0–9 are represented. Remember, we must ignore track 8, the parity track, since this does not form part of the coding system. We must also ignore the vertical stream of *feed holes* that run the entire length of the tape, in between tracks 3 and 4. These are used to physically transport the tape through tape reading and tape punching devices.

Firstly, *all* the digits have a hole punched in tracks 5 and 6. From then on the digits can be read as decimal equivalents of their respective binary punchings. Thus, the digit 1 is equivalent to binary 1 plus holes in tracks 5 and 6. The digit 2 is equivalent to binary 2 plus holes in tracks 5 and 6, digit 3 is equivalent to binary 3 plus holes in tracks 5 and 6, and so on.

The alphabetic characters follow a similar system. *All* the alphabetic characters, A–Z, have a hole punched in track 7. From then on the letters follow an ascending binary count from 1 to 26. For example, the letter A is equivalent to binary 1 plus a hole in track 7. The letter B is equivalent to binary 2 plus a hole in track 7; the letter C is equivalent to binary 3 plus a hole in track 7; right the way up to the letter Z which is equivalent to binary 26 plus a hole in track 7.

The DELETE character has holes punched in *all* tracks since this represents the only way of nullifying every character.

This leaves only 12 characters from the full 50 that perhaps do not lend themselves to easy recognition. It is easy to see why punched tape is so popular when over 75% of the characters can so easily be read by the human operator.

You may have noticed the absence of the decimal point within the ISO character set. The recommendations of BS3635 (sect. 6.3.1) state that the decimal point shall NOT be shown in a control tape and that its position shall be implied by the format of the dimension. However, some systems and data input by other means, for example a keyboard or a host computer, may necessitate that the decimal point character will be required by the control unit.

4.3/3 The EIA standard code

The standard code adopted by the American Standards Association is the **EIA** code. EIA stands for Electronic Industries Association. This code was widely used before the ISO code became established. Nowadays, most CNC control systems will accept punched tape in either format.

The EIA code is also a 7-bit code constructed in an 8-track format. It too utilises a track for parity checks. In this case, however, it is track 5 that is used for this purpose. In a similar way to the ISO code its arrangement facilitates the reading of its various characters, although it is somewhat more cumbersome.

Fig. 4/5 Representation of the EIA numerical control code

DECIMAL EQUIVALENT	b7	b6	b5	b4	b3	b2	b1	NAME OF CHARACTER	CHARACTER SYMBOL	8 (b7)	7 (b6)	6 (b5)	P	4 (b4)	F	3 (b3)	2 (b2)	1 (b1)
0	0	0	0	0	0	0	0	SPACE	SP						•			
1	0	0	0	0	0	0	1		1						•			•
2	0	0	0	0	0	1	0		2						•		•	
3	0	0	0	0	0	1	1		3				•		•		•	•
4	0	0	0	0	1	0	0		4						•	•		
5	0	0	0	0	1	0	1		5				•		•	•		•
6	0	0	0	0	1	1	0		6				•		•	•	•	
7	0	0	0	0	1	1	1		7						•	•	•	•
8	0	0	0	1	0	0	0		8					•	•			
9	0	0	0	1	0	0	1		9				•	•	•			•
11	0	0	0	1	0	1	1	END OF BLOCK	EOB					•	•		•	•
16	0	0	1	0	0	0	0		Ø			•			•			
17	0	0	1	0	0	0	1	OPTIONAL BLOCK SKIP	/			•	•		•			•
18	0	0	1	0	0	1	0		S			•	•		•		•	
19	0	0	1	0	0	1	1		T			•			•		•	•
20	0	0	1	0	1	0	0		U			•	•		•	•		
21	0	0	1	0	1	0	1		V			•			•	•		•
22	0	0	1	0	1	1	0		W			•			•	•	•	
23	0	0	1	0	1	1	1		X			•	•		•	•	•	•
24	0	0	1	1	0	0	0		Y			•	•	•	•			
25	0	0	1	1	0	0	1		Z			•		•	•			•
26	0	0	1	1	0	1	0	BACKSPACE	BS			•		•	•		•	
27	0	0	1	1	0	1	1	COMMA	,			•	•	•	•		•	•
29	0	0	1	1	1	0	1	TABULATE	TAB			•	•	•	•	•		•
32	0	1	0	0	0	0	0	MINUS SIGN	−		•				•			
33	0	1	0	0	0	0	1		J		•		•		•			•
34	0	1	0	0	0	1	0		K		•		•		•		•	
35	0	1	0	0	0	1	1		L		•				•		•	•
36	0	1	0	0	1	0	0		M		•		•		•	•		
37	0	1	0	0	1	0	1		N		•				•	•		•
38	0	1	0	0	1	1	0		O		•				•	•	•	
39	0	1	0	0	1	1	1		P		•		•		•	•	•	•
40	0	1	0	1	0	0	0		Q		•		•	•	•			
41	0	1	0	1	0	0	1		R		•			•	•			•
43	0	1	0	1	0	1	1	PROGRAM START	%		•		•	•	•		•	•
48	0	1	1	0	0	0	0	PLUS SIGN	+		•	•	•		•			
49	0	1	1	0	0	0	1		A		•	•			•			•
50	0	1	1	0	0	1	0		B		•	•			•		•	
51	0	1	1	0	0	1	1		C		•	•	•		•		•	•
52	0	1	1	0	1	0	0		D		•	•			•	•		
53	0	1	1	0	1	0	1		E		•	•	•		•	•		•
54	0	1	1	0	1	1	0		F		•	•	•		•	•	•	
55	0	1	1	0	1	1	1		G		•	•			•	•	•	•
56	0	1	1	1	0	0	0		H		•	•		•	•			
57	0	1	1	1	0	0	1		I		•	•	•	•	•			•
58	0	1	1	1	0	1	0	LOWER CASE			•	•	•	•	•		•	
60	0	1	1	1	1	0	0	UPPER CASE			•	•	•	•	•	•		
63	0	1	1	1	1	1	1	DELETE	DEL		•	•	•	•	•	•	•	•
64	1	0	0	0	0	0	0	CARRIAGE RETURN	CR	•					•			

Fig. 4/5 Representation of the EIA numerical control code

71

The EIA character set is shown in Fig. 4/5, and the following comments will illustrate its make-up.

Once again we must ignore the feed holes in between tracks 3 and 4, and this time, we must also ignore track 5. This is slightly more confusing since this track does not have a "binary weighting" or a decimal equivalent associated with it, and as such, should be ignored when calculating the decimal equivalents of the punchings. The track values are as follows:

8	7	6	5	4	3	2	1	track number
2^6	2^5	2^4	P	2^3	2^2	2^1	2^0	
64	32	16	0	8	4	2	1	decimal value

The digits 1–9 are punched as their values appear in binary. That is, decimal 1 is equivalent to binary 1, decimal 2 is equivalent to binary 2, and so on. The digit 0 has the decimal equivalent 16, i.e. a single hole punched in track 6.

Alphabetic characters are indentified by having holes punched in tracks 6 and/or 7. They are arranged in three sets: letters A to I all have holes punched in tracks 6 *and* 7, letters J to R all have a hole punched in track 7, and letters S to Z all have a hole punched in track 6. In addition, each set follows an ascending binary count. Thus, A and J are equivalent to binary 1 plus their extra hole(s). Similarly, B and K are equivalent to binary 2 plus their extra hole(s), and so on.

Zero is indicated by a single hole punched in track 6 and, as with the ISO code, the DELETE symbol punches a hole in every track.

Note that carriage return is represented by a single hole punched in track 8. This is the only character that uses track eight. It is very useful, therefore, for identifying the different blocks of information encoded on the tape. A new block will start immediately after encountering a hole punched in track 8.

The CNC code systems discussed above will be valid no matter what form of data input is employed. It is merely convenient to use the medium of punched tape to explain their make-up.

Note that the term "byte", used to describe the smallest unit of memory required to store a single character of information, is borrowed from the computer scientists. More formally, 1 byte comprises 8 bits. So each single memory location within the computer is itself made up from eight bits. Each bit is then capable of being **set** (binary 1) or **reset** (binary 0) to form the unique bit patterns of the individual characters.

4.4 Data input

4.4/0 Punched tape

Punched tapes for use in CNC applications are a standard 25 mm (1-inch) wide. They have a capacity for storing 10-characters per 25 mm. Thus, by measuring the length of the tape (in mm), dividing by 25 and multiplying by 10, it is possible to express the length of the part program by the number of characters it contains. Alternatively, the length of a part program may be expressed simply by quoting the length of the tape on which it is stored.

The right-hand edge of the tape is the *reference edge*. This is the edge adjacent to track 1. The offset of the feed holes (between tracks 3 and 4) helps to identify the reference edge. This offset ensures that tapes cannot be loaded into tape readers reverse way round. The direction of the tape can be identified in a number of ways. Many paper tapes have "direction arrows" printed on the upper face of the tape. In addition, when the paper tape is severed from its parent roll, the leading end will be pointed and the trailing end recessed. A typical punched tape is illustrated in Fig. 4/6.

Punched tapes may be read by mechanical, optical or pneumatic readers. The principle of operation of these devices is shown in Fig. 4/7a, b and c.

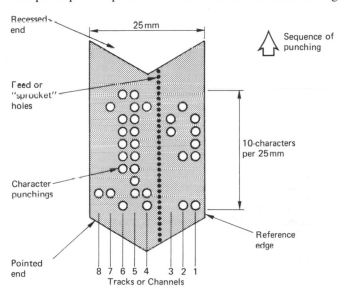

Fig. 4/6 25 mm wide punched tape

a) Mechanical Tape Readers

Mechanical tape readers employ eight spring-loaded wires or fingers corresponding to the tracks, or channels, running across the tape. The fingers bear against the tape as it passes through the reader. When a hole is encountered, the finger momentarily passes through the hole and makes an electrical contact. A voltage is passed, thus registering a binary 1. Punched tape may be read, by mechanical means, at around 50 characters per second. Because of frictional and inertial considerations it is considered a relatively slow means of input.

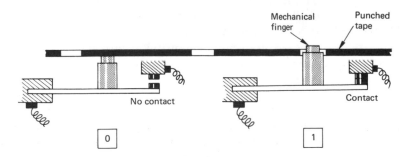

Fig. 4/7a Principle of the mechanical punched tape reader

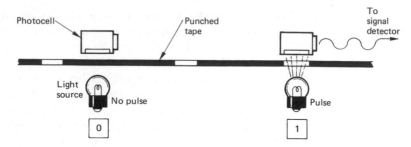

Fig. 4/7b Principle of the optical punched tape reader

Photocell

Punched tape

To signal detector

Light source   No pulse

0

Pulse

1

b) Optical Tape Readers

Optical tape readers employ transducers called *photo-electric cells* or *photo-cells*. A photo-cell is an electronic device which converts light energy into electrical energy. Eight photo-cells are arranged along the width of the tape, one per track. A light source is provided, opposite the bank of photo-cells, which can illuminate any particular cell. The tape is transported in between the photo-cells and the light source. When a hole is encountered, the light shines through the hole and the small beam of light strikes the photo-cell. The photo-cell converts the light dot into an electrical pulse which represents a binary 1. Since the operation is free from mechanical considerations, tape reading speeds can attain 2000 characters per second. For reliable operation, the light source and photo-cells must be kept clean and the punched tape should ideally be made from an opaque medium.

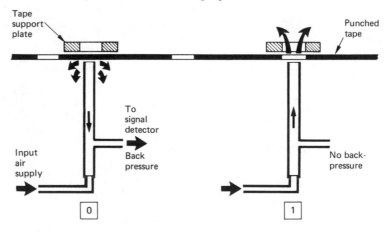

Fig. 4/7c Principle of the pneumatic punched tape reader

Tape support plate

Punched tape

To signal detector

Input air supply   Back pressure

0

No back-pressure

1

c) Pneumatic Tape Readers

Pneumatic readers require a regulated compressed air supply. This air supply is directed through eight specially designed tubes which have two outlets. The first outlet is near the surface of the punched tape (for simplicity we shall refer to this as the main outlet), and the second connected to a signal detector, usually some form of diaphragm actuating device. The punched tape passes over, and in very close proximity to, the main outlet. It is prevented from being "blown away" by a tape support plate which also has holes provided in it to allow the escape of air. When the tape covers the main outlet, the free escape of air is restricted and a back pressure is set up within the supply tube. This back pressure is detected and maintains a stable (binary 0) condition. When a hole in the punched tape uncovers the main outlet, air is allowed to escape freely through the tape and through the tape support plate. The resulting loss of back pressure in the supply tube is sensed by the diaphragm and thus a binary 1 is registered.

All tape readers will require additional circuitry to synchronise read cycles with tape speed and, of course, additional punching devices are required to punch the media.

Although punched tape is often referred to as punched paper tape, a number of alternative materials are now being used which offer more desirable properties—for example, greater mechanical strength to lessen the likelihood of damage by tearing or worn sprocket feed holes, oil and water resistance, opaque and non-reflective properties, etc. Polyester or polyester/paper laminates are common alternatives.

4.4/1   Manual data input (MDI)

Manual data input is the term given to data entry, into the CNC control unit, via the console keyboard. Complete part programs may be entered at the machine by this method. Since this renders the machine idle for the period of time that data is being entered, its use must be considered less efficient than alternative means. There are, however, some machines that permit MDI whilst machining is taking place. The most common application for MDI is the editing of part programs already resident in the controller's memory. This has the added advantage that, once edited, the program may be downloaded (saved) to the backing store or re-punched automatically by outputting to a tape punch.

Individual characters are identified according to the same coding system but their bit patterns are encoded by computer logic.

One variation of MDI that is gaining favour is a concept called **conversational programming**. The CNC machines are programmed via a question and answer technique whereby a resident program "asks" the operator a series of questions. In response to the operator's input, and by accessing a pre-programmed data file, the computer control can:

Select numerical values for use within machining calculations.
Perform calculations to optimise machining conditions.
Identify and configure standard tools with offsets/compensations, etc.
Calculate cutter paths and coordinates.
Generate the part program to machine the component.

A typical dialogue from the machine would ask the operator to identify such things as:

Material to be cut.
Surface roughness tolerance.
Machined shape required.
Size of the raw material blank.
Clearances, machining allowances, cut directions, etc.
Tools and tool details, etc.

The operator may then examine and "prove" the program via computer graphics simulation on the console VDU. The generated part program can be edited or "overruled" by the operator if necessary, and the end product saved on backing store or punched into tape as appropriate.

Although there is some sacrifice in machine utilisation, actual programming time is minimal and much tedious production engineering work (calculating feeds and speeds, etc.) is eliminated. More importantly, perhaps, is the degree of involvement that is returned to the operator. Whilst the introduction of CNC machining techniques has done much for the efficiency of the metal cutting industries, it has also brought about a certain alienation of the machine tool operator. Conversational programming enables the operator to retain and exercise skills rather than inherit the task of becoming merely a "machine minder". CNC, by some, is considered to be a management control. Since industrial personnel relations is predominantly a management concern, concepts such as conversational programming may prove to be significant in the introduction of new technology into what have been traditionally skilled manual functions.

4.4/2 Magnetic tape and disc

Magnetic tapes and discs are becoming more widely applied to CNC applications because of their proven design and usage with computers generally. They are reliable, robust, relatively cheap and can be built into original equipment as extremely neat and compact units.

They both employ the principle of storing data, in coded form, by means of magnetised spots on a magnetic medium. Both magnetic tape and magnetic discs are re-usable in that data may be erased and new data saved. Care must be exercised when handling since dirt, grease or other foreign matter on the magnetic surface may cause data "drop out"; resulting in inconsistent data transfer. Finally, data stored on both these media can be corrupted if they are brought into close proximity with magnets or stray magnetic fields.

Magnetic tape is a cheap and convenient way of storing large volumes of data in a comparatively small space. When the tape is housed in cassette form it is easy to handle, easy to store, and is well protected. A typical cassette tape is 6 mm wide and can store around 100 characters per 25 mm. The data transfer rate will, of course, depend on the speed at which the cassette drive operates. At 100 mm per second, for example, it can transfer up to 400 characters/sec. In computer terms this is relatively slow. Magnetic tape, like punched tape, is a **serial access** medium. That is, to isolate any piece of data on the tape it is necessary to read all the information before it. It is rather like music recorded on a musicassette. It is often quite difficult to pinpoint a particular piece of music accurately. This is because the tape passes the read/write heads in a continuous (serial) manner. Thus, if many part programs are recorded on a single cassette, some inconvenience may be incurred in locating the one required. Since the data is recorded magnetically, it is impossible for an operator to read the contents of the tape. In fact it is impossible, by inspection, to determine whether the tape contains information or not.

Magnetic discs, in contrast, are **random access** devices. That is, any single piece of data recorded on the disc can be accessed as easily, and as quickly, as any other. The more common discs (because of considerations of cost) are the flexible or **floppy discs**. These are circular discs and, like magnetic tape, consist of a plastic material coated with a layer of metal oxide which can be magnetised. The disc is enclosed in a square protective sleeve. The data is stored on concentric tracks which are arranged (electronically) on

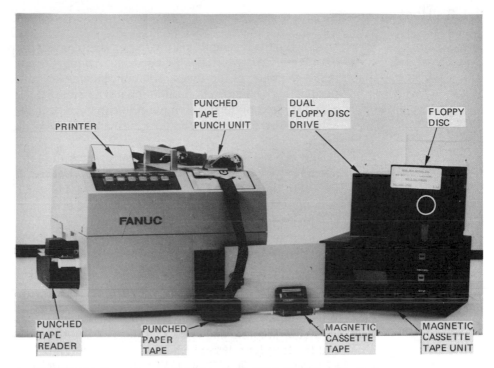

Fig. 4/8 Backing stores and storage media [*Aids Data Systems Ltd*]

the surface of the disc. The standard sizes are 8 inches diameter and 5.25 inches diameter. Unlike magnetic tape, it is possible to use both sides of the disc, for storage, if the correct hardware is available. A read/write head moves across the surface of the disc while it rotates (at about 300 rev/min). In this way it is possible to select just the piece of data required, rather like selecting a single track on an LP record. The capacity of floppy discs may range from 100 K to 1200 K characters. Data transfer rates are considerably faster than for magnetic tape. A typical transfer rate is 20 K characters per second. Typical backing stores and associated storage media are illustrated in Fig. 4/8.

Tape, disc and other peripheral equipment are relatively slow in operation, compared to the central processing unit. It is practice to feed such peripheral devices from a block of random access memory called a **buffer**. The CPU fills this buffer with the information it needs to transfer and can then resume its normal activities. The peripheral can then, simultaneously, carry out the transfer from the buffer. When the buffer is empty, the peripheral device signals the CPU to again fill the buffer. This is called **handshaking** and enables the CPU to continue with its processing unhampered by slow peripheral devices. Incoming data, from peripheral devices, also enters the control unit via the buffer. Buffers are quite separate areas of memory from the main internal memory of the control unit. A typical buffer size would be 512 bytes and a number of buffers may be active at any one time.

4.4/3 Host computer

It is a relatively simple matter to get one computer to communicate with another. (This is further discussed in sections 4.6 and 4.7.) Since transfer is direct, much of the time-consuming inefficiency and unreliability of slow peripherals is eliminated. In addition, it is possible to harness the vast computational abilities of the host computer to carry out complex and time-consuming calculations, and format the data automatically into the form of a part program.

The process of transferring part programs into the memory of a CNC machine tool from a host computer is called **Direct Numerical Control** or DNC. In many large installations the host computer has access to massive data files and may be linked to many different machine tools. As one machine completes its current job, the host computer can arrange for the next part program to be **downloaded**. Thus, a whole manufacturing installation may be under the control of a single master computer.

4.4/4 Error checking and parity

When employing computer technology it is essential that the data being manipulated is accurate and correct. In many cases the amount of data is so vast, and the transfer rates so high, that human operators cannot detect errors due to data transmission. A number of error checking devices have been developed to detect such errors and duly inform the operator. Two such devices are the concept of a *parity check* and the use of *check digits*.

Parity Checking
CNC tape code systems, represented in punched tape, utilise a 7-bit code. An eighth bit is provided as a *parity bit*. The parity bit is a checking device used to determine whether a character has been coded, or transmitted, correctly. The parity bit is set to a 1 or a 0 so that the total number of 1s in the character is even for **even parity** or odd for **odd parity**. For example, the ISO tape code specifies even parity. Reference back to Fig. 4/4 will confirm that the number of holes representing any character (along any row) will be even. If the code for a particular character results in an odd number of holes, then a hole is punched in the parity track, track 8. The EIA system specifies odd parity and uses track 5 as the parity track. Thus, all rows within the EIA tape code will contain an odd number of holes. If the code for any particular character contains an even number of holes, an extra hole is punched in track 5 automatically, by the tape punch. When a tape is read into the CNC control unit, a parity check is made on the incoming data. If a parity fault is detected, transmission will cease and the operator will be informed by some form of error alarm or error message. Common causes of parity faults and tape read errors are shown in Fig. 4/9.

Punched tape is a permanent medium in that the information stored on it cannot easily be changed. With CNC this is not a major drawback. If changes are required to the stored information the tape can be read into the CNC control unit, edited via the console keyboard, and a new (updated) tape punched directly. A problem may arise if a tape has become damaged and tape read errors prevent it from being read into the control unit successfully.

Fig. 4/9 Common causes of tape read errors

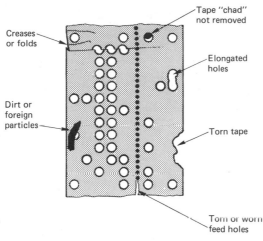

Fig. 4/10 Repair of damaged punched tape

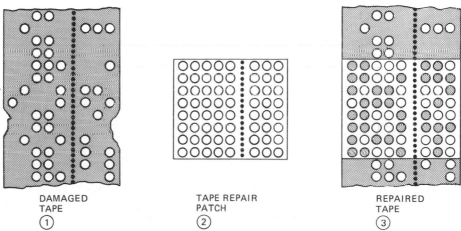

DAMAGED TAPE ①

TAPE REPAIR PATCH ②

REPAIRED TAPE ③

In such cases, tapes may be repaired sufficient for them to be read into the control unit. A common method is to repair the faulty portion of the tape with a pre-punched tape patch. The patches are strong, resilient and usually thinner than the original tape material. They are self-adhesive to provide a quick and efficient repair. Their use is illustrated in Fig. 4/10. This should be seen as a temporary measure to render a damaged tape readable. Once the tape has been successfully read into the control unit, a new tape should be punched at the earliest convenience. Information on punched tapes can be deleted, edited and spliced but the procedure is time-consuming and cumbersome.

4.4/5 Check digits or checksums

Numeric or coded data items of particular importance often have **check digits** (sometimes called **checksums**) attached to them. The value of the check digit is determined from the characters that make up the data item, using a simple formula. The system reading the data uses the same formula to compute the check digit. If the data item has been transmitted correctly then the check digits will be the same; if not, the difference will be detected, transmission aborted, and the operator alerted by an error alarm or error message.

For example, to compute a check digit for the number 12345 we might proceed as follows:

1 Assign each digit a weighting factor:

$$1 \quad 2 \quad 3 \quad 4 \quad 5 \quad \text{data item}$$
$$\vdots \quad \vdots \quad \vdots \quad \vdots \quad \vdots$$
$$6 \quad 5 \quad 4 \quad 3 \quad 2 \quad \text{weighting factor}$$

2 Multiply each digit by its weighting factor and add:

$$(1 \times 6) + (2 \times 5) + (3 \times 4) + (4 \times 3) + (5 \times 2) = 50$$

3 Divide by 11 and note the remainder:

$$50/11 = 4 \text{ remainder } 6$$

4 The check digit is formed by subtracting the remainder from 11:

$$11 - 6 = \mathbf{5}$$

5 The check digit is placed as the least significant digit of the number:

12345**5**

When the number is checked, each digit is assigned the same weighting factor as in **1** and the check digit is assigned a weighting factor of 1:

$$1 \quad 2 \quad 3 \quad 4 \quad 5 \quad 5 \quad \text{data item}$$
$$\vdots \quad \vdots \quad \vdots \quad \vdots \quad \vdots \quad \vdots$$
$$6 \quad 5 \quad 4 \quad 3 \quad 2 \quad 1 \quad \text{weighting factor}$$

Each digit is multiplied by its weighting factor and added:

$$6 + 10 + 12 + 12 + 10 + 5 = 55$$

If the resulting number is exactly divisible by 11 then the test is satisfied.

If alpha characters are present within the code, as will be the case in part programs, then the data items can be dealt with in a number of ways. For example, each alphabetic character can be assigned a value depending on its position within the alphabet (A = 1, B = 2, Z = 26, etc.), or alternatively, each character can be processed on the basis of its decimal equivalent value.

4.4/6 Data transmission

Earlier in this chapter it was stated that computer memory can be visualised as rows of storage locations (likened to individual pigeon holes), which are capable of holding a single character, or byte, of information. Furthermore, each character is represented, in coded form, by eight binary digits or bits. These bits, in computer memory, are in fact electronic switches which could either be switched on (set to binary 1) or switched off (set to binary 0). Thus, it can be inferred that each memory location is, in reality, a set of 8-bits and that a bit being set to 1 is indicated by a small voltage being present at that switch.

When data is being transmitted between a computer device and a peripheral (as with DNC for example), it is done so by sensing (and saving) the voltage level of each individual bit of a particular memory location. These voltage levels are re-constituted, in the same order, and stored by the peripheral device. All data transfer is accomplished in this way.

Obviously the computer device will have to be connected to the peripheral in some way, normally by a cable. A moment's thought will confirm that there are two ways of transferring the voltage, or bit patterns, which represent the data. The first, and perhaps the most obvious way, is to provide each bit with its own wire along which the voltage level can be sensed. This will require a cable comprising 8 wires, one for each bit. Such cable is supplied with the eight wires arranged side by side as a flat ribbon, and is known as *ribbon cable*. Because the voltage level of each bit appears on its wire at the same time as its neighbour, a byte of data can be transmitted in one go. This is known as **parallel transmission**.

The second way in which data transfer can be accomplished is by transmitting each item of data, a bit at a time, down a single cable. Although this will be somewhat slower it has the distinct advantage that only one wire is required, or two if data transfer both ways is required. This may not seem to be that much of an advantage but it means that installations become cheaper and that data can be transferred very easily, over long distances, via existing telephone networks. Thus, local, national and even international exchanges of data can be accomplished with comparative ease. This system of transmitting data is known as **serial transmission**.

4.4/7 The RS232 interface

Although both methods of data transfer described above are widely used within computer systems, the serial transfer system has become standard in the CNC sector. This is because of its compatibility with many major devices within the computer industry itself (large mini and mainframe computers, for example), and the obvious advantages associated with being able to transfer data down existing telephone networks.

If serial transfer is to be successful there has to be some system whereby individual bits can be differentiated, and also where one byte of data stops and the next byte starts. Remember: serial data transmission appears, in practice, as a stream of voltage ONs and OFFs travelling along a single wire. An American (EIA) standard known as the RS232C standard for Serial Data Transmission has been adopted for this purpose. Within this standard there is facility for specifying parameters such as odd, even or no parity checking on transmitted data, the speed at which data is sent (the baud rate), and the format of the synchronising bits (start and stop bits) required to differentiate between different bytes of data.

It has to be arranged that the transfer speed of the transmitting computer is compatible with that of the receiving computer. This is termed the **baud rate** where baud means "bits per second". Transfer rates are usually selectable between 75 and 19200 baud. As a rough guide, baud rate divided by 10 gives the transfer speed in characters per second. It must be stressed that the various settings on the receiving device must be completely identical to those of the transmitting device for successful transfer to occur.

Normally, a collection of electronics can arrange that any output can be made serially. This collection of electronics is often available on a single plug-in circuit board known as an **RS232 interface**. The various parameters, mentioned earlier, may be changed either by software, within a program, or by hardware selection by providing switches or links on the circuit board.

Nearly all CNC control systems support an RS232C communication facility. It may be specified as standard equipment or may need to be specified as a fitted option. It is usually provided to support punched tape equipment (as well as DNC), since few control units are fitted with integral tape punch and read facilities. On some controllers the RS232C interface may be identified as a V24 connection.

Questions 4

1 State *five* principal advantages of employing computer-based systems.
2 Explain the difference between a general-purpose computer, a dedicated computer and a microprocessor.
3 Draw a block diagram of a typical computer system, stating the function of each sub-system.
4 State *three* typical input devices and *three* typical output devices that can be used in computer-based systems applied to engineering.
5 Define the following terms in the context of computer memory devices: RAM, ROM, byte, bit, and backing store.
6 Explain why the binary system is so important in computer-based systems.
7 Convert the following decimal numbers into their binary equivalents: 1, 32, 157, 255, 129, 65, and 5.
8 Convert the following binary numbers into their decimal equivalents: 11001001, 11000001, 01000000, 00000000, 00000001, and 10101010.
9 Punched tape used in CNC applications has 8-tracks. Why was this number of tracks chosen as a standard?
10 Describe *three* ways in which punched tape can be read.
11 Explain how alphabetic characters and other text symbols can be represented on punched tape.
12 Discuss the relative advantages and disadvantages of storing part program data on punched tape and magnetic discs or tapes.
13 Explain the differences between ASCII, ISO and EIA coding systems.
14 Explain how very large and very small dimensions are represented on 8-track punched tape.
15 State *two* error-checking devices that can be used to detect input data errors and explain how they work.
16 State *four* means of inputting part program data into the control unit of a CNC machine.
17 State and explain *three* ways in which the length of a part program can be stated.
18 Explain the difference, outlining any advantages and disadvantages, between serial and parallel data transmission.
19 What is an RS232 interface and where would it be used?
20 What is meant by the term "baud rate" and where would it be used?

Drawing considerations   5

5.1   Drawing implications for CNC

5.1/0   Component design for CNC

It was mentioned in Chapter 1, Section 1.4/1, that CNC will impose a certain influence on the design process. Components should be designed for ease of production when using CNC-related techniques. Thought must also be given, at the design stage, to future developments that could occur within the production environment. For example, the use of automated handling techniques using robots or automatic pallet changing (APC) facilities, for the loading and unloading of components. The introduction of Flexible Manufacturing Systems with the attendant possibility of continuous unmanned working, or the possible changeover to Computer Aided Part Programming (CAPP) techniques.

A number of simple design points may be relevant when designing components for production by CNC methods. Some examples are offered below:

a) Component shapes should be designed such that they can be produced by standard production tooling. The need for form tools should be eliminated.

b) The variety of hole sizes should be kept to a minimum. Each different size hole necessitates at least one tool change, more if pilot holes are required. Each tool change incurs heavy time penalties, and the need for operator intervention if automatic tool changing (ATC) facilities are not available.

c) The number of small-diameter holes should be minimised if at all possible. Small drills and taps break easily!

d) Common component features such as fillet radii, undercuts, grooves, counterbores, chamfers, etc., should be standardised to reduce the variety of tool stocks held, and help reduce the number of tool changes required.

e) Sharp corners should be eliminated on both internal and external forms. The use of radii and fillets corresponding to standard size cutters will simplify part programming and manufacture.

f) Cutter path access should be simplified by providing forms that allow tool run-in and run-out.

g) Machining should be kept to as few planes as possible to avoid excessive component handling.

h) The variation in height of machined surfaces should be kept to a minimum, to reduce lost time in tool movement.

i) If the component is symmetrical, a salient location (and orientation) feature should be provided. This will reduce handling time during set-up and make automated handling easier.

j) If the component is an awkward shape it may be beneficial to provide sacrificial clamping lugs that may be removed at a later stage. This will reduce set-up times. It may also allow the component to be machined, in one setting, without the need for programmed breaks to change clamping positions.

k) It may be relevant to provide a feature that may be grasped for component manipulation. This will allow the easy integration of robotic devices for loading and unloading especially where unmanned machining may be required.

l) The use of stock size raw material should be considered, where appropriate, to reduce the amount of necessary machining.

The above points are not meant to be exhaustive, merely to stimulate thought in the design of components for production by CNC methods. In all cases the designer should ask: "Would I like to make the component to my design?".

5.1/1 Detail drawings for CNC

The majority of components produced by CNC machining techniques are provided to the part programmer in the form of conventional detail drawings. Where the detail drawings are done "in-house", much can be done to aid the part programmer in the production of the part program.

Component drawings produced for manufacture by conventional machining operations often need considerable interpretation by the conventional machine tool operator. Correct production of the part therefore relies on the correct interpretation given to the information on the drawing, by the operator. Such information may not be laid down in standards but is used as a result of common and historical usage. Some examples are shown in Fig. 5/1.

Such instructions, whilst sufficient to convey the relevant information to a skilled and experienced craftsman, are too imprecise to be of value to an automatic CNC machine tool. Many detail drawings are dimensioned with respect to the function of the component. As such they may contain information that is not entirely compatible with production by CNC techniques. Much information will be irrelevant in cases where conversational programming or computer aided part programming techniques are employed. (These techniques are explained in greater detail in Chapter 7.) Similarly, much information may be incomplete if it references other separate components, within say, the same assembly or sub-assembly.

Where the part programmer has to interpret such information, it means duplicating the effort already expended by the detail draughtsman and creating a potential source of misinterpretation and error.

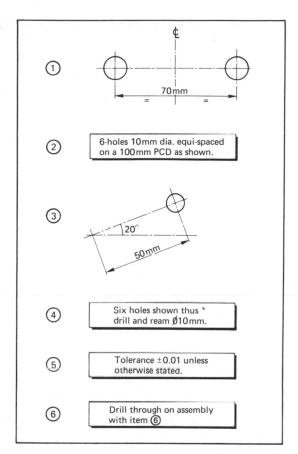

Fig. 5/1 Drawing details incompatible with CNC machining techniques

Detail drawing will only be carried out effectively, at the onset, if the detail draughtsman is fully familiar with the mode of operation and facilities available on the CNC machine in question. The detail draughtsman must be fully conversant with (at least) the type of coordinate systems offered by the machine tool, the system employed for specifying machine datums, and any limitations imposed on the type and/or size of tooling that can be accommodated. The appropriate use of advanced machining facilities (outlined in Chapter 7) may influence the way in which the component is dimensioned. At the fundamental level, for example, it will become unnecessary to apply tolerances to location dimensions between, say, hole centres. The inherent locational accuracy of the machine will be well within any drawing tolerance specified.

5.1/2 Process planning for CNC

The part programmer will often also carry out the task of process planning. Indeed, the part programming/process planning procedures cannot really be divorced if the best results are to be achieved in terms of continuity and efficiency. **Process planning** is the procedure of deciding what operations are to be done on the component, in what order, and with what tooling and workholding facilities.

Both the process planning and part programming aspects of manufacture occur after the detail drawings of a component have been prepared. It is beneficial, however, if the detail draughtsman has a knowledge of the procedure and the thought processes involved in undertaking such tasks.

A typical sequence of events is described below:

1   DECIDE ON COMPONENT DATUM

Consideration to be given to the function of the component, any mating or related parts of an assembly, the form of supply of the material, any previous machining operations, and ease of establishment at the machine tool. Datum planes parallel with machine movements are desirable.

2   PLAN WORKHOLDING

Consideration to be given to component strength and clamping forces, types of clamp, position of clamps for ease of access and operation, maximum amount of machining to be carried out in one setting, clamp change positions, subsequent machining operations, and the possible use of simple fixtures or special chuck/vice jaws. Turning fixtures need special consideration to ensure adequate strength and balancing due to high spindle speeds.

3   PLAN OPERATION SEQUENCE

Attempt to minimise the distance travelled by the spindle during positioning; minimise tool changes and clamp changes, etc. All roughing cuts to be completed before any finishing cuts are attempted. Make maximum use of a single component setting, or tool setting. Similar operations to be grouped in the interests of economy of tool movement and tool changes. Ensure that paths taken by the cutting tools, especially under rapid traverse, will be achieved safely in respect of collisions with clamping arrangements or component features. Programmed breaks to be provided to carry out component inspection, swarf clearance and clamp changes where appropriate.

4   DECIDE ON TOOLING

Cutter types, diameters and lengths to be specified and allocated unique tool numbers. Tool length offset and compensation values to be considered, and a suitable tool change coordinate position decided.

5   TABULATE COORDINATE DIMENSIONS

Dimensions at all tool/cutter change points, hole centres and other relevant features to be identified and tabulated from the established datum. The format of dimensional values to be decided in accordance with the requirements of the machine tool. Computer aided part programming techniques may be considered.

6   ASCERTAIN PART PROGRAM FORMAT

Part program instructions to be considered in relation to the specific format required by the CNC control system.

7   CODE THE PART PROGRAM

The part program code to be formulated according to the format, command word and machining facilities of the specific CNC machine tool. Feed and speed values to be calculated and applied. Elegant and efficient part programming is a skill that is enhanced by experience and by a thorough knowledge and application of the facilities available.

8   CHECK, REVIEW and MODIFY

Dimensional, geometrical and procedural details to be checked and verified. The part program may then have to be modified in the light of other factors.

9 FORMULATE DOCUMENTATION

All documentation to be gathered and brought up to date. Any outstanding documentation to be originated. All documentation is then issued as current, and filed for future reference.

The detail draughtsman can, consciously or otherwise, exert considerable influence on points 1, 2, 4, 5, 7 and 9 above. This influence may, however, be to the assistance or the detriment of the part programmer; steps should be taken, within the organisation, to ensure that it is the former by encouraging a unified approach.

5.2 Datums and coordinate systems

5.2/0 Datums

A **workpiece datum** may be defined as a point, line or surface from which dimensions are referenced. The datum is a fundamental concept in the production of functional working drawings. Students of engineering should be well versed in the choice and importance of datum features.

When considering component manufacture via CNC machining, the concept of the **machine datum** must receive equal importance. The machine datum is an established position, within the programmable area of movement of the machine, about which the machine makes its programmed, dimensional moves. It is often termed the "zero datum" or more commonly simply "zero". Each axis of movement on the machine must have an established **zero datum point**. This zero datum may be fixed (by the machine tool manufacturer) or be user-defined, and set, by the operator. It follows that the machine datum must be referenced, in some way, to the workpiece datum. This will depend on where the component is mounted on the machine.

Machine datum facilities generally take three forms. An individual machine may employ one or more of these facilities.

FIXED ZERO The term **fixed zero** refers to an absolute, fixed point on the machine. The machine tool slides will be traversed to the fixed zero, normally under manual control, before any machining can take place. This manoeuvre resets the axis movement registers to zero. All programmable movements will be related to this point. The fixed zero can be identified, to the control system, by limit switches at the extreme limits of axis movement. On a milling machine, for example, any of the four corners of the worktable may be the fixed zero location (in X and Y). One or other of the left-hand corners is usually chosen.

As the name implies, the machine zero datum is fixed and cannot be repositioned by the operator. In practice, a small amount of manual movement, about the fixed zero, is usually available to assist in positioning the workpiece.

On machines utilising this datum type, an important principle known as **single quadrant positioning** is established. This means that all programmable dimensions are assumed to be positive in sign from the fixed zero. There

Fig. 5/2 Machine datum divides the programmable worktable area into 4 quadrants

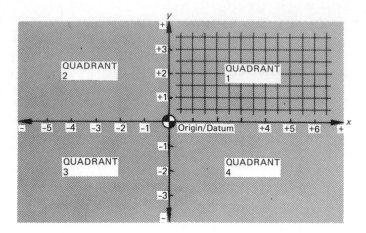

can be no negative dimensions or movements. This, arguably, simplifies the job of the part programmer.

The term "single quadrant positioning" is derived from the representation, on paper, of a two-dimensional graph. The axes of the graph divide the plotting area into four quadrants. They are numbered, by convention, as shown in Fig. 5/2. Consider quadrant 1. All plottable points must be positive in sign. By contrast all plottable points in quadrant 3 must be negative in sign. On some machines the fixed datum may be the rear left-hand corner of the worktable. This suggests that it is quadrant 2 that becomes the programmable area, permitting the use of negative Y-dimensions. This is *not* the case and all programmed dimensions must still be positive in sign. It should be assumed that it is the graphical axes that have been repositioned, to suit the machine tool.

With this system all programmed moves must be made relative to the fixed datum of the machine. This makes the part programming of the component more difficult since the position of the component, on the worktable, must be accurately specified. There will be a strong reliance on workpiece positioning systems, such as fixtures or grid plates, with this datum type.

ZERO or DATUM SHIFT   This facility allows the machine tool zero point to be moved (or "shifted") to any desired position within the programmable area of movement. **Datum shift** permits the convenient location of the workpiece on the machine worktable since the machine zero can now be made to correspond with the workpiece datum. Part programming is simplified since the component can be positioned on the worktable, at the convenience of the machine tool operator.

Although the zero point can be shifted to any convenient point, the system operates in the same way as that for the fixed zero machines. There can be no negative dimensions or movements. Movement is restricted to the quadrant having the zero point as its origin. This effectively restricts the available programmable movement available.

FLOATING ZERO   The above facilities were available on many early NC machines. The **floating zero** or *full floating zero* is by far the most common datum system in operation on modern CNC machines.

This system allows the zero point to be positioned at any convenient location within the programmable area of the machine. Any coordinate position may be described, numerically, in terms of positive or negative dimensions from the zero point. This system effectively places the machine datum at the origin of the conventional two-dimensional graphical coordinate system. Machining is allowed in all four quadrants.

This system is by far the most convenient since it allows the detail draughtsman to position the component datum, for convenience, at the time of drawing production. This is the more natural approach since, for example, the component datum may be the centre of a hole. This could not be accommodated using the fixed zero or datum shift systems.

The above discussion has concentrated on datums in a single plane of movement, X–Y or X–Z. On CNC milling machines there is a third axis of movement. If the machine is a 3C machine, then the above comments for full floating zero will apply. In the case of a 2CL machine, the zero position of the spindle will normally be the position at which the quill is fully retracted. (The terms 3C and 2CL are explained in Chapter 3, section 3.3/3.) In all cases the machine will be positioned, under manual control, at the desired location and the axis registers set to read zero.

5.2/1 Coordinate systems

Fundamentally there are three types of coordinate system that may be employed by a control system to position the tool or cutter in relation to the workpiece. Each have their application and may be used independently or mixed, according to the features present within the component. Great versatility is afforded to the designer, detail draughtsman and part programmer in that either system may be selected, and re-selected, from within a part program. In addition, both inch and metric units can also be selected from within a part program thus allowing the two systems to be freely mixed, if desired.

ABSOLUTE COORDINATES **Absolute coordinates** are dimensions which are always measured from the same datum position. This will be a datum position in the X-axis, the Y-axis and the Z-axis. The datum position for each axis is user definable and may or may not be the same point.

Absolute dimensions utilise a system of X, Y coordinates (often called *Rectangular or Cartesian coordinates*), which are all referenced from the datum position. Each coordinate position must be identified by both an X and a Y dimension (X and Z when turning), if working in two dimensions. This is the system employed in conventional graph plotting applications. When three-dimensional applications are considered, as will be the case when milling, each coordinate must be identified by an X-dimension, a Y-dimension and a Z-dimension.

This is often the preferred system of coordinates since there is no accumulation, or build-up, of tolerances between individual dimensions. An example of absolute coordinate dimensioning is illustrated in Fig. 5/3a.

INCREMENTAL COORDINATES **Incremental coordinates** are related by the end point of the previous dimension. The point just dimensioned (or

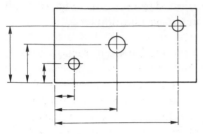

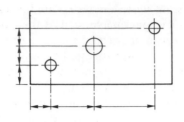

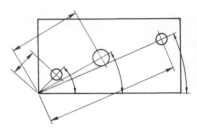

Fig. 5/3a Absolute
coordinates

Fig. 5/3b Incremental
coordinates

Fig. 5/3c Polar
coordinates

visited, when machining) serves as the datum for the next point (or positional move). Incremental coordinates are illustrated in Fig. 5/3b.

This system has previously been discouraged since a build-up of tolerances can occur between the different points or features dimensioned. This means that some features could be in error when manufactured, or require tighter tolerances to be specified in order that specific relationships can be maintained. This is less of a problem with CNC machining techniques since the inherent accuracy and repeatability of the machine tools themselves diminish these problems. Indeed for some CNC machining applications, the system of incremental coordinates is preferred. Common repeatable patterns (hole patterns or pockets), or symmetrical patterns (mirror image features), benefit from being dimensioned using incremental coordinates. Advanced machining techniques (discussed in Chapter 7) employing incremental coordinates will considerably simplify part program production. In addition, many component features will be made re-locatable. A re-locatable feature is one that only has to be described once (within a part program) and can then be reproduced many times, at different positions, within the same component.

POLAR COORDINATES **Polar coordinates** describe the position of a feature by a length, measured from a specified point, and an angle (measured in degrees) from a specified datum axis. This system is illustrated in Fig. 5/3c.

Most CNC machine tools operate only in linear dimensions. Thus, any features shown on a drawing using polar coordinates will have to be converted into either absolute or incremental rectangular coordinates. This will involve the application of trigonometry, Pythagoras' Theorem, and/or the sine and cosine rules. In general, polar coordinates are not recommended unless special machining options, available within the CNC control, require dimensional information in this form.

One such common option which may require the use of polar coordinates is that of a number of equi-spaced holes on a pitch circle diameter (PCD). It may only be necessary to specify the PCD, the number of holes required, the coordinates of the PCD centre, and the position of the first hole given in polar coordinates.

A second (similar) facility that may be available is the ability to "rotate" a feature about a specified point of rotation. The production of components comprising a series of spoke-like features can be produced very easily using such facilities. More importantly, if the detail draughtsman is aware of these facilities, component drawings can be dimensioned to take advantage of them and thus save considerable draughting and part programming time.

The choice of coordinate system will be determined by many considerations: the function of the component, the relationship of the component with other components in the same assembly, inspection techniques employed for checking, the presence of repeated features, ease of manufacture, location of the datum feature, and so on. In many cases the datum will itself be a feature of the component. There is no reason why it should not be at a point outside the profile of the component. This will often be the case on components that make up assemblies or sub-assemblies.

Most CNC machines have the capability to work in all three coordinate systems and can switch systems in mid-program. Whilst adopting a single system of coordinates (preferably absolute) will serve to simplify programming (and dimensioning), the features of the CNC machine should be fully understood. The system applicable to the situation should be employed. Much time and duplication of effort may be saved.

5.3   Documentation for CNC

5.3/0   The importance of documentation

The purpose of documentation associated with CNC machining is fundamentally fourfold. The provision of complete and up-to-date documentation can make useful contributions far beyond the immediate requirements of enabling the production of components.

1   *Communication*

Many concepts, ideas and procedures have to be organised and communicated by various people, to various people, during the production life cycle of a manufactured component. Much information may be factual and much may be creative or original. Information relating to component drawings, machine utilisation, tooling, workholding and setting, operation sequences, modifications and updates, even the part program tape itself, constitute documentation.

In all cases the priority should be that all information is communicated in as accurate, unambiguous, clear and efficient way as possible. For these reasons most of the common documents associated with CNC are pre-printed standard-format forms or sheets. The exact make-up of these standard forms will be individual to the particular organisation or machine tool. Their use should be encouraged. They save time and effort in putting information down on paper, they enforce a logical, structured and uncluttered presentation of information, and the format is such that it is familiar to all interested parties within the organisation.

It is important to document information that is both required and potentially useful, either currently or at a later date.

2   *Factual Information*

Factual information is constantly required, for various purposes, by many different people. Factual information provides evidence of what has (or what has not) actually been done. It can thus assist in tracing errors or technical difficulties, providing current status information, answering queries and enquiries from customers or internal departments, providing information concerning manufacturing costs, and so on. No organisation can function without factual information. Indeed, in many cases, contractual and legal obligations make it absolutely necessary that certain information is retained, even for a period of years after the components have been produced. Verbal communications, in conversation or by telephone, are not a reliable means of communication in a production environment. Many people in various departments need to be aware of all decisions likely to affect them. Verbal information can easily be forgotten, may not be passed on, and is thus unreliable.

Before production commences there can often be a considerable time period after the planning, estimating or tendering for particular jobs. Alternatively, considerable periods of time can elapse between production runs of the same component. It is essential that through maintaining complete and accurate factual information, working familiarity is regained in as short a time as possible.

3 *Historical Data*

Historical data, or records as they are more commonly referred to, can be a source of much benefit if they are well organised. They can provide a wealth of experience, and time savings, when estimating and planning for the production of similar components.

The success (or otherwise) of previously used manufacturing methods may provide useful technical, managerial or accountancy planning information. For example, new versions of the same or similar components may affect manufacturing processes, technical or other changes may render tools and/or operations obsolete, and so on. They will be needed for a responsive treatment to spares production, and product information enquiries, often some years after original manufacture.

Of course, the success, or otherwise, of historical records is dependent solely on the accuracy, integrity and completeness of the data they hold, and the ability to locate it from many different avenues of enquiry.

4 *Calculations*

Errors in part programs can be attributed in many cases to erroneous calculations made when determining the cutter path coordinates. In order to quickly isolate the source of such errors, and prevent much duplication, all calculations should be carefully documented. This must be done in a clear manner such that anyone having cause to check any calculations can do so with minimum difficulty.

The importance of documentation may only become apparent when it becomes conspicuous by its absence. By this time it is usually too late for the purpose for which it is required. Prudent and conscientious maintenance of documentation will do much to ensure that tasks are accomplished smoothly and efficiently.

5.3/1 Types of documentation

The following types of documentation represent the minimum required for CNC machining operations. There is no standard format for such documents. It is sufficient that all the required information is provided in a clear, concise and unambiguous manner. There should be clearly defined areas of responsibility for the provision of associated documentation. Remember it is often down to the machine tool operator to pull the many strings together to produce a successful component. It is essential that the operator has access to all relevant information in order to avoid waste of time and having an expensive machine tool sitting idle. Decisions during machining require that all relevant information is to hand. More importantly, all the documentation must be current.

1 *Component Drawing*

Many people will need access to a fully dimensioned component drawing. This should be drawn to scale although not necessarily full size. However, missing dimensions should never be scaled using a rule. The component datum should be clearly indicated on the machine tool operators copy.

2 *Tabulated Coordinated Dimensions*

The cutter path is really a series of coordinate points. A separate coordinate point will be specified where the cutter needs to change direction in response to a programmed command. They may be referred to as *cutter change points*. It is the responsibility of the part programmer to determine the change points based on the particular operation sequence specified. Hole centre locations and tool change positions are further examples of such change points.

It is good practice that these change points should be uniquely identified, numbered and written down in tabular form. The part programmer will undoubtedly have to calculate these points during compilation of the part program. It is very little extra effort to write them down in a table of coordinates. This table will prove an invaluable aid: for example, to the operator when setting and proving the part program at the machine, as a check to the part programmer that all component features have been accounted for, and for locating and editing errors when they occur.

3 *Part Program Listing*

In support of the above items, a printed listing of the part program (generated from the part program tape) is essential when editing at the machine is likely. The operator will undoubtedly have access to digital or alphanumeric readout facilities at the machine console. However, this can be tedious and time-consuming to read and search, especially if the part program is long. It will also enable the operator to verify the actual operation of the part program, at the machine.

4 *Operation Sheet*

The practical aspects of producing the component at the machine are set down on a document called an operation sheet. It may also be termed an *operation schedule*. The operation sheet itemises each separate operation in chronological order. It specifies the operation, and identifies the tools required for that operation together with any setting details. It also states cutting speeds and feeds (even though these will be "inbuilt" within the program), the means of workholding, clamping and setting, and any other anticipated requirements.

The intention is to remove the bulk of the thinking and planning time away from the machine.

The important term is "anticipated requirements". Decisions taken at the planning and part programming stage may not truly reflect the actual conditions on the shop floor, at the time of manufacture. There can be many reasons for this, many quite unforseeable. If information regarding the original intentions is available, it is obviously much easier and quicker to make alternative arrangements to secure successful manufacture. A typical operation sheet is shown in Fig. 5/4.

Fig. 5/4 A typical operation sheet

OPERATION SCHEDULE						SHEET	OF	
COMPONENT DESC.						MACHINE		
PART No.								
JOB No.								
PREPARED BY				DEPARTMENT		ISSUE DATE		
OP. No.	TOOL No.	NOMINAL SIZE AND TOOL DESCRIPTION	CUTTING SPEED	FEED RATE	SPINDLE SPEED	OPERATION		WORK HOLDING

Fig. 5/5 A typical tooling sheet

TOOL PREPARATION DATA					SHEET	OF	
COMPONENT DESC.					MACHINE		
PART No.							
JOB No.							
PREPARED BY			DEPARTMENT		ISSUED DATE		
TOOL No.	TOOL OFFSET No.	NOMINAL SIZE AND TOOL DESCRIPTION	TOOL OFFSET		CUTTER DIAMETER OR NOSE RADIUS COMPENSATION	INSERT TYPE	HOLDER TYPE
			X	Z			

5 Tooling Sheet

This item can take different forms. Firstly, it may simply be a list of all the tooling required to manufacture a particular component. The tools may be specified generally, by type and size, or specifically by unique identification. The latter will be the case when a properly organised tool stores is in operation, where automatic tool changing facilities are in operation, or when pre-set tooling techniques are employed. Each tool needs to be allocated an individual tool number. This identifies which tool or turret station the tool must be loaded into or in what sequence the tool is to be used within the part program.

Tooling sheets may also specify details relating to tool offset and compensation values. In such cases the dimensional offset or compensation value will be specified along with the designated tool number and offset register number. These will be referenced from within the part program during tool changes. The operator will be required to edit these values into the relevant offset and compensation registers, prior to commencing machining. An example of a typical tooling sheet is shown in Fig. 5/5.

6 *Workholding Details*

Methods of workholding, clamping arrangements, clamping positions, and setting instructions need to be communicated to the machine tool operator. Special milling or turning fixtures, clamping devices, chuck or vice jaws, etc., are identified where appropriate. Datums need to be identified. Where special setting aids such as grid plates or gauges are employed, location and setting information is provided.

In many cases the provision of workholding and setting details is, perhaps, better conveyed (or at least supplemented) by simple but effective sketches, showing the required details.

5.3/2 Identification of documentation

ALL documentation, regardless of type, should contain basic but essential identification.

a) The *job* to which the document relates should be clearly identified: by stating the component description, an in-house or customer part number, or an in-house or customer job number. If the document refers to a particular operation, this too should be clearly indicated.

b) The *date or status of issue*: by simply stating the date of issue or by providing an issue number or code. In some cases an originally issued document is continually updated by writing in any modifications (together with a date or modification number), in a designated place reserved for this purpose. Unless there is a tight control on the number and destination of all issued documents, this system can be unreliable.

c) If the document is *machine or equipment specific*, this should be made clear by identifying the machine and/or equipment together with any acceptable alternatives.

d) The *originator* and the originating department of the document should be clearly indicated. All queries may then be followed up at source. Signatures alone can often be somewhat illegible!

e) If the document is part of a larger set, the *sheet number* and the total number of sheets in the set should be indicated. The accepted convention is to include a statement which reads "SHEET 1 OF 3" or something similar. This ensures that any incomplete documentation is highlighted.

The provision of the above identification on all documents will ensure that misinterpretation is minimised and accuracy is maximised in all such communication.

5.4 Calculations

5.4/0 The importance of calculations

Part programming involves describing the cutter path to be traversed in coordinate terms. The description of the cutter path is a relatively straightforward task. The identification of the actual coordinates can be more involved. In all cases, the ability to read engineering drawings must be matched by the ability to extract the required information from the drawing. Invariably this requires calculations and the correct application of some common mathematical and geometrical principles. If these principles are mastered then the writing of part programs will be accomplished with ease and with relative speed. Both factors are important in a competitive environment.

If calculations are in error, then the resulting part program, and ultimately the machined component, will also be in error. If an error is detected, the task of isolating the error then has to be undertaken. This will involve recalculation and a duplication of the work already carried out. The situation will be further hampered if this has to be carried out by a different person to that who formulated the original part program. It is essential, therefore, that the approach to all calculations should be accompanied by the following attributes:

Meticulous and thorough attention to detail.
A logical and methodical approach.
Care and accuracy.
Clear and ordered working (fully documented on paper).
Careful identification and filing for future reference.

The common calculations required are not difficult, but require adequate (and honest!) practice. Time spent in mastering the basics will be amply repaid. A sound working knowledge of arithmetic, transposition of formulae, simple algebra and basic geometry must be acquired.

The following sections outline the common additional mathematical techniques required to augment the above.

5.4/1 Pythagoras' Theorem

Pythagoras' Theorem concerns the lengths of the sides of *right-angled triangles*. In coordinate calculations it is often possible to construct a right-angled triangle between three points. If the lengths of any two sides of the triangle are known (or can be calculated easily), then this theorem can be used to determine the length of the third side.

The theorem states that if we take the lengths of the two smaller sides of a right-angle triangle, square them, and add them together, the result will be equal to the length of the hypotenuse (the longest side) when it is squared.

Consider the triangle in Fig. 5/6. The theorem can be expressed in mathematical terms as:

$$AB^2 = AC^2 + BC^2$$

Fig. 5/6 Right-angled triangle

or expressed another way:

$$AB = \sqrt[2]{AC^2 + BC^2}.$$

By simple transposition, these formulae can be re-arranged to give expressions for side AC when sides AB and BC are known, and side BC when sides AB and AC are known. Carry out this exercise before studying the following example.

EXAMPLE   In the triangle of Fig. 5/6, if side AB = 100 mm and side AC = 20 mm, calculate the length of side BC.

$AB^2 = AC^2 + BC^2$	state the formula
$BC^2 = AB^2 - AC^2$	step by step carefully re-arrange to isolate the unknown side
$BC = \sqrt[2]{AB^2 - AC^2}$	
$\quad = \sqrt[2]{100^2 - 20^2}$	substitute known values
$\quad = \sqrt[2]{10\,000 - 400}$	carefully perform calculations
$\quad = \sqrt[2]{9600}$	
$BC = 97.98\,\text{mm}$	state answer with correct units

5.4/2   Trigonometry

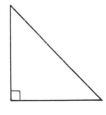

1.

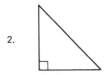

2.

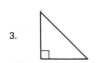

3.

Fig. 5/7 Three similar right-angled triangles

Trigonometry is also involved with the relationships of right-angled triangles. This time, the angles as well as the sides are considered.

Consider the three triangles shown in Fig. 5/7. The triangles are similar. In each case the angles are the same; it is the size (length of the sides) that is different.

EXERCISE   In triangle number 1, measure the length of the base and the length of the longest side, the hypotenuse. Divide the length of the base by the length of the hypotenuse and note down the answer. Repeat the above for the two other triangles. The results are the same.

Repeat the above, this time dividing the vertical side by the hypotenuse. In each case the result is the same.

Repeat the above for a third time and divide the vertical side by the base. As before, the results are the same.

The ratios of the lengths of the sides in similar right-angled triangles are always the same, provided the angles in each of the triangles are the same. They are called "trigonometrical" or 'trig.' ratios. Since they are the same no matter what the size of the triangle, they can be calculated and put into tables. These are the familiar trig. tables. When using trig. tables the ratio of the measured sides can be used to determine the angle, or, conversely, the angle can be used to determine the ratio of the measured sides. In the latter case, if the ratio and one side is known, then the length of the other sides can be determined.

Trigonometrical ratios are thus dependent on the angles in the triangle. In order to be able to differentiate which sides and which angles are being considered, a special notation and language is used. This is illustrated in Fig. 5/8.

Fig. 5/8
Trigonometrical ratios

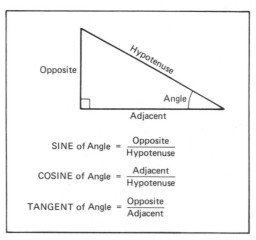

$$\text{SINE of Angle} = \frac{\text{Opposite}}{\text{Hypotenuse}}$$

$$\text{COSINE of Angle} = \frac{\text{Adjacent}}{\text{Hypotenuse}}$$

$$\text{TANGENT of Angle} = \frac{\text{Opposite}}{\text{Adjacent}}$$

The trig. ratios, in conjunction with a set of trigonometrical tables (or a suitable calculator), can be used in solving problems involving coordinate dimension. The use of trig. ratios is limited to right-angled triangles only.

Consider the triangle in the previous example, in Fig. 5/6. The angle ABC can be found by applying the Sine ratio as follows:

$$\text{Sin} = \frac{\text{opposite}}{\text{hypotenuse}}$$ state formula

$$\text{Sin (angle ABC)} = \frac{20}{100}$$ substitute known values

$$= 0.2$$ carefully perform calculation

angle ABC = arc Sin (0.2) re-arrange to find angle

angle ABC = 11 degrees 32 minutes read value from sine tables

5.4/3   The sine rule

Both Pythagoras' Theorem and the use of trig. ratios dictate that a right-angled triangle must first be identified. There are many instances where triangles can be formed but they are not right-angled triangles. In such cases it is possible to use a formula called the Sine Rule. Although it uses the Sine trig. ratio it can be applied to all triangles.

The proof of why it works lies in geometry and the application of algebra. It is not important since we only require to know how to use it. The formula states:

$$\frac{a}{\sin A} = \frac{b}{\sin B} = \frac{c}{\sin C}$$

where a is the length of the side *opposite* angle A, etc.

It may seem rather strange that a formula can have 2 "equals" signs. In actual fact it cannot. This is merely a notation that says each term is equal to any other term. In practice, just two terms (any two) would be used together.

Fig. 5/9 Hole positions requiring solution by the sine rule

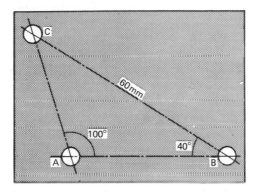

EXAMPLE Consider the hole positions shown in Fig. 5/9. Calculate the length AB.

angle $C = 180° - (100° + 40°) = 40°$ angles in a triangle $= 180°$

$$\frac{a}{\sin A} = \frac{c}{\sin C}$$ from the sine rule

In terms of Fig. 5/9,

$$\frac{BC}{\sin A} = \frac{AB}{\sin C}$$ sine rule applied to example

$$AB = \sin C \times \frac{BC}{\sin A}$$ re-arrange for side AB

$$= \sin 40° \times \frac{60}{\sin 100°}$$ substitute known values

$$= 0.6428 \times \frac{60}{0.9848}$$ look up values in sine tables

$$= 0.6428 \times 60.9260$$ carefully perform calculations

$$AB = 39.163 \text{ mm}$$ state answer with correct units

It is left as an exercise to calculate the length of side AC.

This highlights the problem of finding (from tables) the sine of an angle greater than 90°. Tables only accommodate values for angles between 0 and 90° since the values simply repeat for other angles. In such cases the angle to be looked up can be obtained from the following simple arithmetic:

For angles greater than 90°   look up (180° − angle).

5.5 Geometrical tolerances

5.5/0 Why geometrical tolerances?

The concept of **dimensional tolerances** is fundamental to the manufacturing process. Dimensional tolerances are really statements of how much inaccuracy can be tolerated in the manufacture of component features. Engineers will be familiar with the principles of applying dimensional tolerances to limit the manufactured sizes of component features. Dimensional tolerances are required because it is impossible to produce anything to an "exact" size.

Dimensional tolerances also restrict, to some extent, errors in the shape, or form, of component features. There are, however, many instances in which dimensions and tolerances of size (however well exerted) cannot guarantee accuracy of form. As a graphic demonstration of this, measure the "diameter" of the familiar British 50p or 20p coin, at a number of positions, using a workshop micrometer; although the "diameter" appears to be constant, the shape is far from round. Other, less graphic examples are illustrated in Fig. 5/10.

Fig. 5/10 Dimensional tolerances alone cannot guarantee accuracy of form

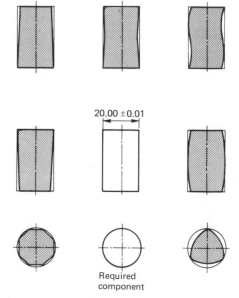

20.00 ±0.01

Required component

There is a clear need for some kind of "shape tolerance" to limit the amount of geometric inaccuracy in component shape and form. Such tolerances are known as **geometrical tolerances** and are the subject of BS308 Part III. Geometrical tolerances specify the maximum allowable variation in form, or position, from true geometry.

Geometrical tolerances should be applied for all requirements which are critical to function or interchangeability. However, if the machinery and techniques used in the production of components can be relied upon to produce the required standard, then geometrical tolerances need not be specified. CNC machining applications generate curves by a series of discrete steps via circular interpolation. Geometrical tolerances may thus have particular relevance when dimensioning curved features.

5.5/1 Geometrical tolerance symbols

A geometrical tolerance is either the width or the diameter of a **tolerance zone** within which a surface, or axis of a hole or cylinder, can lie. The resulting part must also satisfy any other conditions of dimensional accuracy or surface texture requirements imposed on it.

Geometrical tolerances are represented, on engineering drawings, by means of standard symbols. These are shown in Fig. 5/11.

The symbols are placed within a rectangular **tolerance frame** together with other information required to specify the tolerance. Additional information must include the size of the tolerance zone and any datum features about which the tolerance is specified. In the interests of clarity, the information

Fig. 5/11 Geometrical tolerance symbols and their interpretations

SYMBOL	CHARACTERISTIC	APPLICATION
——	STRAIGHTNESS	The indicated feature must lie between two parallel lines, at the specified distance apart, parallel to the datum axis or plane.
▱	FLATNESS	The indicated feature must lie between two parallel planes at the specified distance apart.
○	ROUNDNESS	The indicated feature must lie within two circles, concentric with each other, a radial distance apart as specified.
⌀̸	CYLINDRICITY	The surface of the feature is required to lie between two cylindrical surfaces, co-axial with each other a specified radial distance apart.
⌒	PROFILE OF A LINE	The indicated profile is required to lie between two lines which envelope a series of circles, having the specified diameter, with their centres on the theoretically correct profile.
⌓	PROFILE OF A SURFACE	The indicated surface is required to lie between two surfaces which envelope a series of spheres, having the specified diameter, with their centres on the geometrically correct surface.
//	PARALLELISM	The indicated features must lie within two parallel straight lines at the specified distance apart.
⊥	SQUARENESS	The indicated feature is required to lie within two planes, at the specified distance apart, which are perpendicular to the specified datum.
∠	ANGULARITY	The axis of the indicated feature must lie within two parallel straight lines, at the specified distance apart, inclined at the angle indicated.
⊕	POSITION	Point of intersection of the indicated feature must lie within a circle of specified diameter, with its centre at the true point of intersection.
◎	CONCENTRICITY	The centre of the indicated feature must lie within a circle, of the specified diameter, concentric with the centre of the indicated datum.
≡	SYMMETRY	The indicated feature must lie within two parallel straight planes, at the specified distance apart, symmetrically disposed about the plane situated symmetrically between the indicated datums.
➚	RUN-OUT	Axial run-out not to exceed the stated value when measured parallel to a specified datum axis.
Ⓜ	MAXIMUM MATERIAL CONDITION	An increase in the specified geometrical tolerance may be permitted, equal to the difference between the MMC limit of size and the actual size.
▭	TRUE POSITION	Any tolerance in position is accommodated by other specified geometrical tolerances.

Fig. 5/12 Geometrical tolerance frames

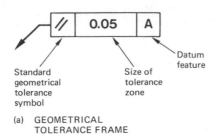

(a)   GEOMETRICAL
       TOLERANCE FRAME

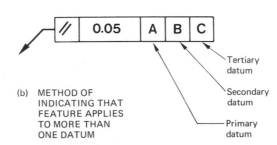

(b)   METHOD OF
       INDICATING THAT
       FEATURE APPLIES
       TO MORE THAN
       ONE DATUM

is presented, within the tolerance frame, in an agreed format. A typical tolerance frame is illustrated in Fig. 5/12.

The tolerance frame is divided into compartments. The extreme left-hand compartment contains the geometrical tolerance symbol. Adjacent to this, moving rightwards, the next compartment holds the size of the tolerance zone. The size of the tolerance zone will be expressed in the same units as those used on the drawing for linear dimensions. The extreme right-hand compartment(s) specify the datum(s) about which the geometrical tolerance applies.

5.5/2   Datum features

Where geometrical tolerances relate to **specific datums**, these datums must be identified on the drawing. These datums may be completely different from the machining datums specified for use by the part programmer. The datum feature is indicated by a leader line from the tolerance frame, terminating in a solid triangle placed to identify the datum feature on the drawing. The datum is identified by an upper case (capital) letter, which is placed in the third compartment of the tolerance frame. This is shown in Fig. 5/13.

Where the tolerance applies to more than one datum, these are indicated in priority order, in additional compartments added on to the right of the tolerance frame. The order in which the datums are specified, reading from left to right, indicate the relative importance of the datum features in priority order.

From left to right	Priority 1	Primary datum	
	Priority 2	Secondary datum	
	Priority 3	Tertiary datum	etc.

If the tolerance frame itself cannot be connected to a datum feature in a clear and simple manner, then a capital letter in its own frame can be used. This must identify the datum feature, by terminating in a solid triangle, in exactly the same way as that used by the standard tolerance frame.

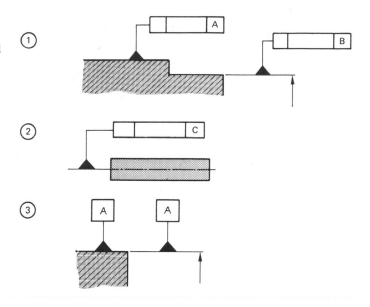

Fig. 5/13 Methods of indicating datum features for geometrical tolerances

5.5/3 Principles of geometrical tolerancing

When specifying geometrical tolerances on working drawings the following general principles will apply:

a) Unless otherwise specified, a geometrical tolerance applies to the whole length, or surface, of the feature concerned.

b) Certain types of geometrical tolerance will require a feature, or features, other than that being toleranced, to be specified as datum(s).

c) The use of a feature as a datum implies that it should itself have adequate accuracy of form or position. This includes any temporary datums that may be used during manufacture. For example, tooling holes, registers or spigots.

d) An element may be of any form, or take up any position within the tolerance zone specified, except where a further restriction is imposed.

e) It may be that one type of geometrical tolerance will automatically limit other types of geometrical error. For example, specifying a parallelism tolerance will automatically limit errors of flatness.

f) The system of indicating geometrical tolerances does not imply the use of any particular method of manufacture, inspection or gauging.

g) In general, geometrical tolerances are to be observed regardless of the actual size of the finished feature. The exception to this is the application of a condition known as **maximum material condition** (MMC).

For further information on the application of geometrical tolerances, and more detail regarding MMC and a condition known as *true position tolerancing*, the reader should consult specialist textbooks or refer to BS308 Part III.

Questions 5

1 Discuss, giving examples, why the adoption of CNC machining techniques will have an effect on component design.

2 Why is it important that the designer and draughtsman should be familiar with the CNC machine tools likely to be used in the manufacture of components?

3 State the steps required to properly plan component production by CNC techniques.

4 Explain the terms "fixed zero", "datum shift" and "floating zero" in the context of CNC machine tool datums.

5 Illustrate, with clear, neat diagrams, the difference between absolute, incremental and polar coordinates. Give an example of where each coordinate system would be preferred.

6 What factors will influence the choice of a component datum?

7 Give *four* reasons why documentation is important when planning manufacture by CNC techniques.

8 State *six* types of documentation, stating their purpose.

9 Explain *five* means of identification that should accompany all documentation.

10 What information should appear on a tooling sheet?

11 What information should appear on an operation sheet?

12 State and explain in your own words, giving examples, Pythagoras' Theorem and when it would be employed.

13 State and explain in your own words, giving examples, the trig. ratios Sine, Cosine and Tangent and when they would be employed.

14 State and explain in your own words, giving examples, the Sine Rule, and when it would be employed.

15 Explain, using neat sketches, why dimensional tolerances alone are insufficient to guarantee accuracy of form.

16 Explain, with reference to standard symbols, what geometrical tolerances are.

17 How are datum features identified when referencing geometrical tolerances?

18 State the principles of applying geometrical tolerances on working drawings.

19 What is single quadrant positioning and what is its main advantage?

20 What procedure should be followed by the machine operator having discovered that dimensional information is incorrect?

Machining considerations 6

6.1 Tooling for CNC

6.1/0 Cutting tool materials

One of the most important qualities that a cutting tool must possess is that *it retains its hardness at the high temperatures generated during the cutting process.* Research into tool material technology has introduced successively better cutting tool materials that can sustain prolonged cutting at more elevated temperatures without loss of hardness.

The evolution of these cutting tool materials is charted in Fig. 6/1. It can be seen that a machining operation that took 100 minutes to complete with a carbon steel tool (around the year 1900) can be accomplished in under a minute using modern multi-coated carbide materials, developed in the 1980s.

The most common cutting tool materials used in CNC applications are *HSS*, *sintered carbides* (notably tungsten carbide), *ceramics*, *CBN* and *polycrystalline diamond*.

Fig. 6/1 Evolution of cutting tool materials [*Sandvik UK Ltd*]

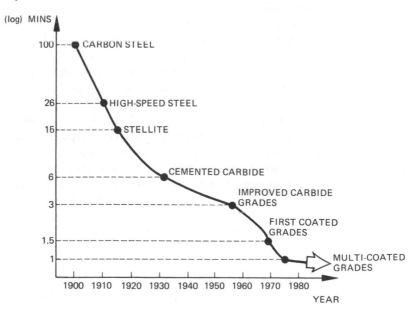

HSS is an alloy of steel containing between 5% and 20% tungsten. It is a popular cutting tool material especially for small-diameter drills and milling cutters. Although very tough, HSS loses its hardness at around 650°C. This limits its effectiveness for CNC applications where high cutting speeds (and attendant higher cutting temperatures) are encountered. Developments in coated HSS cutting tool materials offer inherent toughness and the lower power cutting characteristics of plain HSS. Machining performance is increased by hard-coating techniques and improved hardening and tempering processes. Chemical processes allow a surface coating of hard metallic nitrides to be deposited on the HSS surface. Titanium nitride coated HSS tools are one such example.

Sintered carbides are by far the most popular cutting tool materials for CNC applications. The performance of sintered carbides is derived from the inherent hardness of the main constituents used. Tungsten (the most popular), titanium and tantalum carbides are all extremely hard. They are also very brittle and less tough than HSS. This means that they cannot normally be used in their pure unprocessed form. The carbides are manufactured in powder form and mixed with other powdered constituents such as cobalt. The powder is then pressed into preformed shapes and sintered. Sintering involves subjecting the preforms to extremely high temperature causing the materials to fuse into a dense and non-porous structure. The cobalt, as well as imparting toughness, acts to cement the carbide grains into the resulting structure. For this reason they are also known as **cemented carbides**. This technique of powder metallurgy allows powdered constituents to be blended together in precisely mixed proportions in a manner not possible by other means. It therefore allows certain properties to be built-in to the final material specification.

In general, a better quality surface finish is obtained when using sintered carbides since higher cutting speeds can be achieved and there is less tendency for chip welding to occur on the top face of the tool. It is common for small deposits of swarf to become welded to the top surface of HSS tools causing what is known as a *built-up edge*. When this brittle built-up edge breaks away from the tool surface, it takes with it some of the tool material. This weakens the tool point and is a common cause of tool failure in HSS tools.

Sintered carbides are more brittle than HSS. For this reason they have to be more rigidly supported. Most carbide tool tips are mounted to present a *negative rake angle* when cutting and this accounts for the higher driving forces required when utilising sintered carbide tooling. Positive and negative rake tool angles are illustrated in Fig. 6/2.

The majority of sintered carbide cutting tool materials being used for CNC turning applications are of the coated type. Uncoated varieties are used predominantly for milling applications because of the intermittent cutting characteristics. **Coated carbides** are so called because the sintered carbide core is further coated with additional wear-resistant carbide coatings. These coatings are "grown" onto the surface of the carbide core since adhesion is achieved by atomic bonding. Single or multi-coated carbide grades are available. Coating materials (having a finer grain structure) include titanium carbide, aluminium oxide and titanium nitride. Tool life in excess of five times that of plain sintered carbide materials can be achieved.

The international standard ISO 513:1975 sets out a classification for the various grades of sintered carbide materials and their applications. There

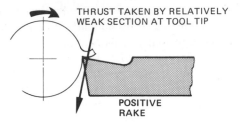

Fig. 6/2 Positive and negative rake tool angles

THRUST TAKEN BY RELATIVELY WEAK SECTION AT TOOL TIP

POSITIVE RAKE

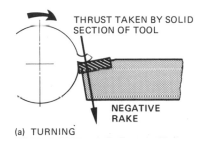

THRUST TAKEN BY SOLID SECTION OF TOOL

NEGATIVE RAKE

(a) TURNING

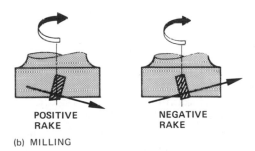

POSITIVE RAKE

NEGATIVE RAKE

(b) MILLING

is no equivalent British Standard at the time of writing. An interpretation of this classification is shown in Fig. 6/3.

Ceramic cutting tool materials are formed from sintered aluminium oxides. They can be used at higher cutting speeds than sintered carbides and exhibit better wear resistance. This is primarily due to the lower temperatures generated at the tool tip. They are not recommended for intermittent cutting (either turning or milling) which probably accounts for their reduced popularity. There is no need to use coolants when employing ceramic cutting tool materials other than to cool the workpiece. Cutting speeds in excess of two and three times that of sintered carbides can easily be achieved using ceramics.

Mixed ceramics are being developed which exhibit somewhat better toughness characteristics. One example is based on silicon nitride and is known as *Sialon*. The name is coined from an anagram of the letters of the main constituents, i.e. SiN (silicon nitride) and AlO (aluminium oxide).

Diamond is an exceptionally hard, naturally occurring material with extremely good wear-resistance properties. It resists compression forces twice as well as tungsten carbide and expands very little with heat. For these reasons it can hold very close tolerances during machining and produces an extremely high-quality surface finish. It finds best application in precision machining and finishing operations. When machining ferrous components, chemical reaction at high cutting temperatures can cause diamond to revert to its original graphite form. For this reason the use of diamond cutting tools is limited to the machining of non-ferrous and non-metallic materials. Synthetic diamond, comprising super-hard constituents, is most commonly employed for CNC applications. The material is then known as *polycrystalline diamond* because of its characteristic structure.

Cubic boron nitride (CBN) is an exciting new cutting tool material second only to diamond in hardness. It is twice as hard as tungsten carbide. In contrast to diamond, CBN is also an extremely tough and stable material that can be used for the machining of very hard ferrous materials. Tool steels, including HSS itself, can be machined efficiently with CBN, and it can withstand the high shock loading caused by interrupted cutting applications. Typically, CBN can be used, without coolant, at two to three times the speeds and feeds associated with ceramic materials. CBN is supplied in insert form. It should be used in clamp-type tool holders, supported by ground tungsten carbide shims, exhibiting negative rake characteristics. It has a structure similar to that of polycrystalline diamond and may thus also be referred to as *polycrystalline cubic boron nitride* (PCBN).

Hardness and wear resistance are not the only factors that should be considered when discussing cutting tool materials. *Rigidity* and the *ability to withstand interrupted cutting* are also important. Rigidity affects the accuracy of the finished component. Maintaining straight, perpendicular and accurate machined features depends on rigid cutters. Carbide is approximately three times more rigid than steel.

Interrupted cutting is not only encountered in milling operations. It can occur as a result of eccentric components, the presence of holes or slots, hard spots in workpiece materials, blow holes or sand inclusions in castings, eccentric circular components, and so on. These characteristics can be encountered during any machining operation and it is essential that the cutting tool materials can withstand the mechanical shocks so produced.

6.1/1 Hardmetal insert tooling

The development of sintered carbide, ceramic and CBN cutting tool materials has significantly reshaped tooling configurations in general. Whilst HSS cutting tools combine cutting edges, shank and retaining features in a solid one-piece tool, sintered carbides (CBN and ceramics) are supplied in the form of small cutting tool tips or inserts. They are known as **hardmetal inserts**.

The inserts are available in a variety of standard shapes and sizes. There is usually more than one cutting edge on each insert. A square or rectangular insert, for example, may have up to 8 cutting edges, a triangular insert up to 6 cutting edges, and so on. When a cutting edge is exhausted, the insert is simply indexed round to expose the next cutting edge. For this reason they may also be referred to as **indexable inserts**. The inserts have to be located, supported and secured by a special toolholder, shank or cartridge before they can be used. Although this may at first sight seem very cumbersome, there are significant advantages to be gained by using inserts.

a) Worn or damaged tools can be indexed (or replaced) very quickly, thus maintaining smoother production and keeping machine downtime to a minimum.

ISO CATEGORY AND COLOUR	BROAD WORKPIECE MATERIAL USAGE	CODE	APPLICATION	CHARACTERISTIC
P (BLUE)	FERROUS METALS WITH LONG CHIPS	P01	Finish turning/boring under vibration-free conditions.	INCREASING CUTTING SPEED / INCREASING WEAR RESISTANCE ↑
		P10	Turning, threading and milling. Light roughing and finishing.	
		P20	Turning, threading and milling. Average cutting conditions, moderate speeds.	
		P30	Turning/milling will withstand heavy feeds and shocks.	
		P40	Turning and milling of castings and stainless steels in unfavourable conditions.	
		P50	Machining under demanding and unfavourable cutting conditions requiring very tough carbide.	
M (YELLOW)	FERROUS METALS WITH LONG OR SHORT CHIPS	M10	Finish-turning alloy steels with light feeds and moderate speeds.	
		M20	Turning/milling of alloy steels. Excellent universal grade.	
	NON-FERROUS METALS	M30	Roughing of difficult-to-machine alloys.	INCREASING FEED / INCREASING TOUGHNESS ↓
		M40	Turning, parting off and roughing of mild steel, non-ferrous metals and light alloys.	
K (RED)	FERROUS METALS WITH SHORT CHIPS	K01	Finish-turning of hard materials and abrasive plastics. Wear-resistant.	
	NON-FERROUS METALS	K10	Turning/milling/drilling of malleable cast iron and non-ferrous alloys.	
	NON METALLIC MATERIALS	K20	Turning/boring/milling of non-ferrous alloys requiring tough carbide.	
		K30	Rough-turning/milling under unfavourable cutting conditions.	
		K40	Turning/milling of non-ferrous materials under unfavourable conditions.	

Fig. 6/3 ISO grading classification for sintered carbides

b) No regrinding is necessary; indeed it is often not possible to successfully grind insert materials. When the insert is fully expended it is discarded. This eliminates the need for expensive re-grinding facilities. Hardmetal inserts can also (although incorrectly) be referred to as **throwaway tips**. In practice, spent cutting tool tips can be "salvaged" and re-used in heavy wear application areas such as mining and quarrying.

c) Correct cutting tool geometry and dimensional accuracy of the inserts are guaranteed by the sintering process. This means that greater use can be made of pre-set tooling techniques. Chip breaker grooves can be formed within the insert to provide inherent swarf control at no extra cost. A chip breaker groove causes the chip to curl and break into short sections, thus making the swarf easier to dispose of.

d) A wide variety of standard shapes and sizes of inserts are available off the shelf. Most machining applications are thus accommodated without the need to provide specialist cutters.

e) The inserts can be employed in milling, turning and drilling toolholders alike. This allows rationalisation and standardisation of tooling to be achieved, keeping tooling costs to a minimum.

ISO 1832 (BS4193 Part 1) sets out a standard specification for the designation of hardmetal inserts for cutting tools. The classification comprises a 10-symbol code. The first 7 symbols are compulsory, the following 2 symbols are optional, and the last symbol (always preceded by a / character) is for the manufacturer's use. An interpretation of the code is shown in Fig. 6/4.

ISO 5608 (BS4193 Part 6) sets out a standard specification for the designation of toolholders and cartridges for use with hardmetal inserts. A **cartridge** is an insert-holding device that can be incorporated into a special cutting tool body (such as a multi-diameter boring bar). The cartridge has facility for both axial and radial adjustment. Inserts used in milling cutter bodies are located in insert "seats". The classification comprises a 10-symbol code. The first 9 symbols are compulsory and the last one should be used when appropriate. An interpretation of the code is shown in Fig. 6/5.

The use of inserts can be applied to almost any style of cutting tool. Turning tools, drills, end mills, peripheral and face mills, slot mills, boring and threading tools can all be configured using inserts. A range of applications is illustrated in Figs. 6/6, 6/7 and 6/8.

For smaller-diameter cutting tools such as drills, end mills and slot drills, etc., below 15 mm diameter, brazed tip tools can be used or solid HSS tools employed. It becomes impractical to apply inserts to tools of such small dimensions.

6.1/2   Choosing hardmetal tooling

There appears to be a bewildering array of insert shapes and sizes, toolholder designs and insert materials to choose from. The success of the machining operation depends, to a large extent, on the final combination employed. The following points need to be considered when making the final choice.

1   Insert clamping system   Inserts may be secured in the toolholder either by wedge-type arrangements or by screwed clamps. Wedge, screw and pin

INSERT SHAPE	NORMAL CLEARANCE ANGLE		TOLERANCE CLASS			FIXING AND CHIP BREAKER	INSERT SIZE	INSERT THICKNESS	INSERT CORNER RADIUS	CUTTING EDGE CONDITION	CUTTING DIRECTION
1	2		3			4	5	6	7	8	9

1 Insert Shape	2 Normal Clearance Angle	3 Tolerance Class	4 Fixing / Chip Breaker	5–7 Size/Thickness/Radius	8 Cutting Edge	9 Cutting Direction
H	A — 3°	A: m ±0.005, s ±0.025, d ±0.025	N	Symbol is the numerical value of the size/thickness/radius.	F	R
O	B — 5°	F: m ±0.005, s ±0.025, d ±0.013	A	Ignore decimal values.	E	L
P	C — 7°	C: m ±0.013, s ±0.025, d ±0.025	R	Single-digit values must be preceded by a 0.	T	N
S	D — 15°	H: m ±0.013, s ±0.025, d ±0.013	M		S	
T	E — 20°	E: m ±0.025, s ±0.025, d ±0.025	F	Example — SQUARE INSERT		
C	F — 25°	G: m ±0.025, s ±0.13, d ±0.025	G	Side length = 16.5 mm, Thickness = 3.18 mm, Corner rad = 0.8 mm		
D	G — 30°	J: m ±0.005, s ±0.025, d ±0.05/±0.13	X			
E	N — 0°	K: m ±0.013, s ±0.025, d ±0.05/±0.13	OTHER			
M	P — 11°	L: m ±0.025, s ±0.025, d ±0.05/±0.13				
V	O — OTHER	M: m ±0.08/±0.18, s ±0.13, d ±0.05/±0.13				
W		U: m ±0.13/±0.38, s ±0.13, d ±0.08/±0.25				
L						
A						
B						
K						
R						

DESIGNATION

16	03	08
5	6	7

EXAMPLE OF ISO INDEXABLE INSERT DESIGNATION

L	P	C	A	12	03	08	F	R
1	2	3	4	5	6	7	8	9

Fig. 6/4 ISO specification for the designation of hardmetal inserts

INSERT CLAMPING SYSTEM	INSERT SHAPE	TOOL STYLE		NORMAL CLEARANCE ANGLE		CUTTING DIRECTION	TOOL HEIGHT	TOOL WIDTH	TOOL LENGTH		INSERT SIZE	QUALIFIED SHANKS
1	2	3		4		5	6	7	8		9	10
C	H	A	B	A	3°	R	Symbol is the numerical value of the height/width.		A	32	Ditto 6 and 7	B
M	O	C	D	B	5°	L	Ignore decimal values.		B	40		F
P	P	E	F	C	7°	N	Single-digit values must be preceded by a 0.		C	50		Q
S	S	G	J	D	15°		For a tool cartridge, precede the symbol with letter C.		D	60		
	T	K	L	E	20°				E	70		
	C	M	N	F	25°				F	80		
	D	R	S	G	30°				G	90		
	E	T	U	N	0°				H	100		
	M	V	W	P	11°				J	110		
	V			O	OTHER				K	125		
	W								L	140		
	L								M	150		
	A								N	160		
	B								P	170		
	K								Q	180		
	R								R	200		
									S	250		
									T	300		
									U	350		
									V	400		
									W	450		
									X	AS STATED		
									Y	500		

EXAMPLE OF ISO TURNING TOOL DESIGNATION

M	S	A	C	R	32	32	L	12	Q
1	2	3	4	5	6	7	8	9	10

Fig. 6/5 ISO specification for the designation of toolholders

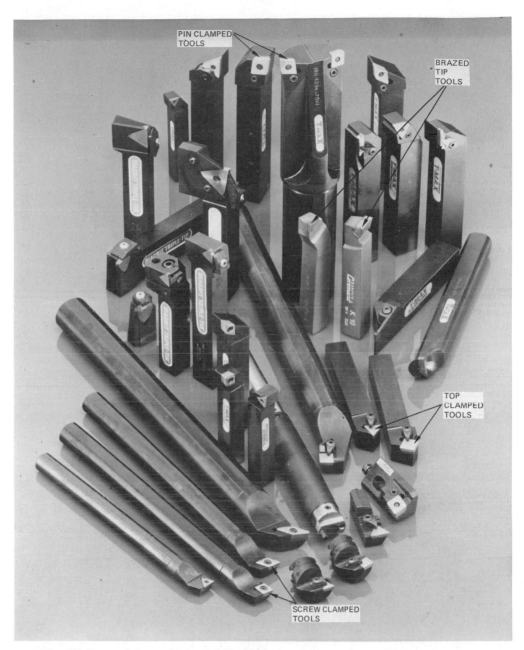

Fig. 6/6 Inserted tip turning tools [*Sandvik*]

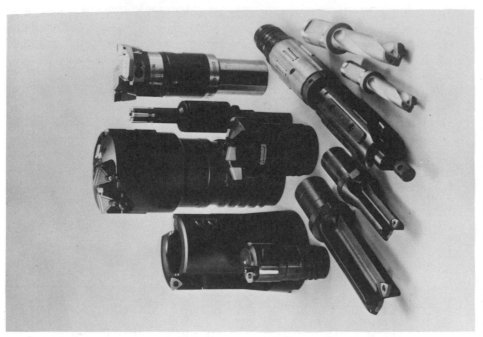

Fig. 6/8 Inserted tip drilling tools [*Sandvik*]

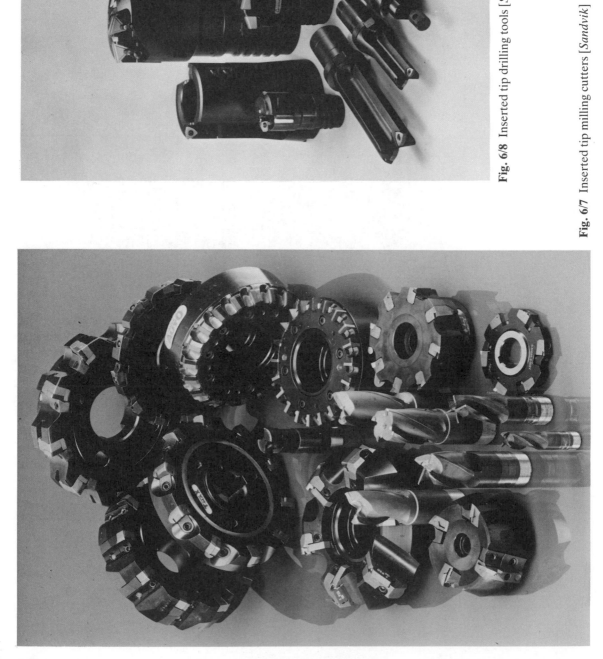

Fig. 6/7 Inserted tip milling cutters [*Sandvik*]

type clamps offer a plain top surface for unhampered chip flow, maximum accessibility and maximum rigidity. The plain top surface allows for maximum accessibility to machined contours. Top clamped inserts offer robust and stable clamping where the direction of cutting forces may have to be considered and where loose mechanical chipbreakers need to be employed.

The four principal clamping systems are illustated in Fig. 6/9.

Although tungsten carbide inserts can be brazed to mild steel shanks, this eliminates the indexing properties of the inserts and is only employcd on small-diameter cutters. Ceramic inserts cannot be brazed (other than by special techniques), since they are non-metallic. They can however be bonded by

Fig. 6/9 Insert clamping systems

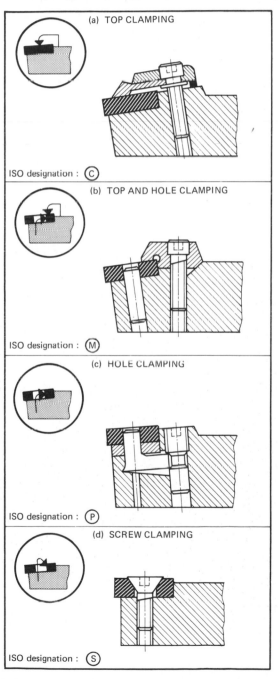

(a) TOP CLAMPING

ISO designation : Ⓒ

(b) TOP AND HOLE CLAMPING

ISO designation : Ⓜ

(c) HOLE CLAMPING

ISO designation : Ⓟ

(d) SCREW CLAMPING

ISO designation : Ⓢ

adhesives. Most commonly they are clamped using top clamp systems since these can distribute the clamping forces over a wide area of the insert.

2 Toolholder type The type of toolholder is dictated, to a large extent, by the shape of the workpiece to be machined. The type of machining operation, whether performing internal or external machining, the operation sequence, tool access considerations, etc., all influence the final choice.

3 Insert type The insert itself is chosen according to a number of factors. The insert shape is determined by the operation to be performed, the toolholder employed, and the contour shapes required on the component. Features such as radii, grooves, chamfers, threads, etc., can all be accommodated by standard inserts. In general, round inserts are stronger in use than square inserts. Square inserts are stronger than triangular inserts, but the latter are more versatile.

Other factors such as chipbreaking ability and capacity, cutting geometry, whether cutting will be continuous or intermittent, and vibration tendencies must also be considered.

4 Insert size The actual size of the insert depends on such factors as the maximum depth of cut, length of cutting edge required, roughing or finish machining (roughing would require a large nose radius), and likely cutting forces. In general, the maximum depth of cut should not exceed two thirds the cutting edge length.

5 Insert grade The grade of insert depends on the workpiece material. The choice of insert grade also determines machining data and the speeds and feeds to be used.

Most cutting tool suppliers have excellent guides for the selection of tools, insert types and insert grades for most machining applications. Comprehensive and specific machining data, for most cutting conditions, is also provided. It is advisable to become familiar with such selection guides and make extensive use of such data when planning component manufacture.

6.1/3 Chipbreakers

One very important aspect of hardmetal inserts is the ability to specify formed-in **chipbreakers**. These are shaped grooves that cause the swarf to curl and break up into chips of manageable size. Chipbreakers can be formed in a variety of other ways. For example, they may be formed by an insert top clamping arrangement or ground in on brazed tip cutters.

Many ductile workpiece materials cause swarf to be produced in long continuous ribbons. This is undesirable in CNC applications for the following reasons:

a) It can seriously hamper the cutting operation by wrapping itself around the workpiece or cutter.

b) It can be a danger to the operator.

c) Much of the heat generated during cutting is accumulated within the swarf and this causes an undesirable build-up of heat that can cause thermal distortion of both workpiece and machine tool structure.

d) In unmanned operation there will be no operator to monitor and remove accumulations of swarf.

Form-sintered chipbreakers have a chipbreaking capacity dependent on the depth of cut and feed rate employed. If an insert is chosen without regard to chipbreaker capacity, then recommended feed-rate/depth-of-cut values will produce poor chip control. Manufacturers can supply details of how chipbreaker capacity can be determined.

The choice of insert should not be made without due consideration of chipbreaking ability and capacity.

6.1/4 Non-insert tooling

Wherever possible, for serious CNC metal cutting applications, insert tooling should be adopted. In cases where solid HSS tooling is employed, the following points are offered to assist in achieving the best possible performance.

The correct *cutting fluid* should always be used. Increased tool life, improved surface finish, more accurate control of size and assisted swarf removal are all benefits that can be realised by the choice and use of the correct fluid. The choice will be based on such factors as tool material, work material and the machining operation. Manufacturers recommendations should be followed.

In general, neat or soluble (water-diluted) oils are available.

Neat oils are preferred when:

a) Cutting operations require anti weld properties.
b) Performing severe operations or machining difficult alloys such as nimonic or heat-resisting alloys.
c) High levels of accuracy and surface finish are required.
d) Machine tool designs require the lubrication afforded by neat oils.
e) The risk of corrosion by water-based fluids must be avoided.

Soluble oils are preferred when:

a) A high degree of cooling is required (e.g. when employing high metal removal rates).
b) There is a potential fire hazard.
c) An economic coolant is required and machine tools can tolerate water-based products.

Tool overhang should be kept to an absolute minimum consistent with the machining operation. Rigidity, and hence tool deflection due to dynamic loading, is proportional to length by a factor of (length)^3. Thus, a drill, end mill or boring bar which is 3 times longer than it should be (because of excessive overhang) will be 27 times less rigid, or exhibit 27 times more deflection under dynamic loading whilst cutting.

End mills should be chosen with the minimum flute length consistent with the machining operation. Excessive flute length contributes to tool "chatter" which can significantly affect tool life. Tool life can be considerably extended by using end mills with the shortest possible flute length.

Accurate machining relies on *accurate location* of the cutting tools. Cleanliness when assembling tools and cutters is of course essential. Some cutting tools, especially HSS drills and end mills, have information regarding the size of the tool stamped on their shank. This stamping causes the deformed material around the characters to stand slightly proud of the shank. These raised burrs can cause lack of concentricity and run-out, when the tool is mounted. Tools with etched details are preferred or, alternatively, the burrs must be carefully removed with a smooth oilstone.

Tool point geometry on conventional twist drills may require modification. In general, centre drills (a combination drill and countersink) are not usually employed on CNC machining centres. Speeds and feeds are generally not compatible with the small drill point and countersink diameters. This means that the drill will not have the benefit of the initial location, provided by the centre drill, sufficient to start its cut. Conventional twist drills have a large web thickness, a blunt chisel point, and an included angle of 118°. These features make it difficult for the drill to start its own hole without wandering off-centre prior to penetrating the material. The blunt point contributes to "rubbing" of the workpiece surface which can cause local work hardening and the drill point to wander off centre.

Web thinning may be carried out to reduce the chisel edge thickness. Care must be taken to ensure tht the resulting chisel point remains central and that too much weakening of the drill point (by excessive grinding) does not take place.

The drill point angle may be modified in order to suit the material being cut. In general, the softer the material to be cut, the more acute (sharp) must be taken to ensure that the resulting chisel point remains central and operation with a conventional "spot drill" or using a slot drill, especially when flat-bottomed holes are required.

For producing short length holes, a hardmetal insert drill should be considered. This is a specially designed drilling tool employing hardmetal inserts. They are available in sizes ranging from about 17 mm to 56 mm in diameter. Two inserts set asymmetrically at the tool tip are placed such that no chisel edge is presented to the work. This balances the resultant cutting forces. The short length of the drill, and its inherent rigidity, prevents wandering at the start of the drilling operation. Such drills can only be used for drilling holes up to ($2 \times$ diameter) deep. There are often no helical flutes, so the drill is inherently strong and rigid. Swarf (in the form of chips) is transported down parallel flutes running the length of the drill. Because the flutes are parallel, swarf does not flow naturally out of the hole as it does with helical flutes. For this reason swarf removal is considerably assisted by cutting fluid. Pressurised cutting fluid is fed direct to the cutting tool tip via a small hole that passes through the drill body. This forces the chips to travel up the drill flutes and out of the hole. The drills are designed for high chip removal rates and it is essential that chip breakage occurs. To utilise them, the machine tools must meet certain requirements in terms of power, spindle speed, feed and cutting fluid pressure. They may be used as a rotating or non-rotating cutting tool. No centre drilled or pre-drilled pilot holes are required. These features enable short hole drilling operations to be carried out easily and quickly without fear of drill breakage or without the need to provide centre drilled or pre-drilled pilot holes.

Alternatively, brazed-carbide/coated-carbide tipped drills may be employed for holes less than 17 mm in diameter. These are essentially "throw-away drills" since regrinding is inadvisable. Best performance of these drills is accomplished using coolant with EP (extreme pressure) additives. Some modern brazed tip drills can be ground with an S-shaped "chisel" edge which improves cutting characteristics. The drills are self-centring and may be used for drilling holes up to (3.5 × diameter) deep.

Typical inserted and brazed tip drilling tools are shown in Fig. 6/8.

6.1/5 Qualified and pre-set tooling

Tool changing and tool setting can be greatly speeded up by the use of qualified tooling and pre-setting techniques.

Qualified tools are tools on which the position of the cutting edge is guaranteed to within close limits of accuracy. The accuracy of the cutting tool tip is quoted with respect to specified datums on the toolholder. Typical position accuracy is within ±0.08 mm. The qualified dimensions are applied to the tool tip from up to three datums. Usually the datums are formed by the toolholder or cartridge. Since the dimensions of the toolholder or cartridge will be known, precise replacement of the tools in the machine tool is possible. The use of hardmetal inserts (and their precise dimensions and geometry) is ideally suited to qualified tooling arrangements. Properly applied qualified tooling can eliminate the need for tool measurement and setting either at or away from the machine.

Turning toolholders are available which are located by three adjustable buttons set into the shank. The buttons can be adjusted (by screwing them in or out) to a qualified dimension. Such tools are termed *semi-qualified tools*. They have the advantage that they can be set to slightly different qualified dimensions enabling them to be used on a variety of machine tools. They do, however, have to be checked and maintained regularly to ensure that the qualified dimensions remain correct. Examples of qualified and semi-qualified turning tools are shown in Fig. 6/10a and b.

One drawback of setting tools at the machine is that the machine has to stand idle and is rendered non-productive. To eliminate this downtime it is possible, by duplicating the tool-mounting arrangements of the machine tool, to set the tools away from the machine. Tool set-ups can be planned and carried out in advance, ensuring continuity of production and minimising downtime due to tool set-ups from job to job. Special-purpose **pre-setting**

Fig. 6/10a Qualified turning tools

Fig. 6/10b Semi-qualified turning tool

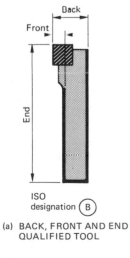

ISO designation (B)

(a) BACK, FRONT AND END QUALIFIED TOOL

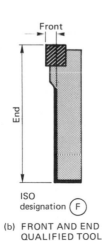

ISO designation (F)

(b) FRONT AND END QUALIFIED TOOL

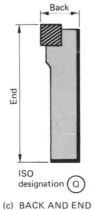

ISO designation (Q)

(c) BACK AND END QUALIFIED TOOL

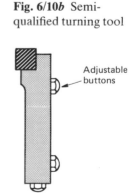

Adjustable buttons

119

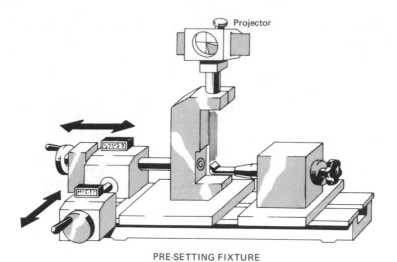

PRE-SETTING FIXTURE

INTERCHANGEABLE
SUB-PLATE

Fig. 6/11 Pre-setting
fixture

fixtures are often used for this purpose. In order that precise tool location can be achieved, the machine datums must also be duplicated on the pre-setting fixture. It is essential that there is facility for accurately setting the tool point for position in three dimensions. To achieve this, pre-setting equipment incorporating precision slideways, digital read-outs and projection equipment offering magnification and angular adjustments is often employed. Interchangeable support plates duplicating different tool mounting details can be used such that different machine tools can be pre-set using the same equipment.

One example of a pre-setting fixture is illustrated in Fig. 6/11. Such a fixture would be used for the setting of turning tools and, with the appropriate mounting plate, tool lengths for milling related cutters.

6.1/6 Tooling systems

Most machining centres provide a standard spindle nose taper for locating tool shanks and toolholders. Tool retention is provided either by hydraulic or quick-release mechanical arrangements depending on whether manual or automatic tool changing facilities are employed.

Most turning centres employ a standard tool-retaining system, within the turret, that both locates and clamps the toolholders.

This is an acceptable state of affairs when dealing with only single machine tools. Standardisation of tooling and pre-setting techniques becomes cumbersome if different machines adopt different standards, and there is a large variety of different types. Many machine tool and cutting tool manufacturers are now offering integrated systems of tool location and clamping that can be used on a variety of different machine tools. Such systems are known as **tooling systems**.

Tooling systems usually comprise a master holder which can be obtained to suit the standard mountings of most machine tools. A range of common adapters and spacers can then be used to mount any cutting tool on any machine. As a result, standardisation of tooling and reduced tool stocks can be achieved. The adaptors are built to high standards of locational accuracy and are usually of a quick-release design. Tool changeover time between jobs is kept to a minimum using these systems. When inserts have worn or broken cutting edges, it is often quicker to replace one adapter with another, rather than attempting to change (or index) an insert at the machine. An insert may take around 6 minutes to replace. Another important advantage of such systems is that they lend themselves to unmanned operation. A turning centre can be attended to by a robot due to the standard size and shape of the adaptors. This is an important factor when considering unmanned operation.

A turning centre incorporating a 10–12 station turret may be limited to short periods of unmanned operation, because of the limited range of tools it can accommodate at any one time. This can be overcome by incorporating automatic tool changing facilities that will swap a tool resident in the turret for a tool in an external tool magazine. A robot in conjunction with a bank of different tools, located at the side of the machine, could increase its capacity for unmanned operation. Both these solutions are, however, expensive.

A typical tooling system concept for a turning centre is shown in Fig. 6/12.

Fig. 6/12 Tooling system for a CNC turning centre

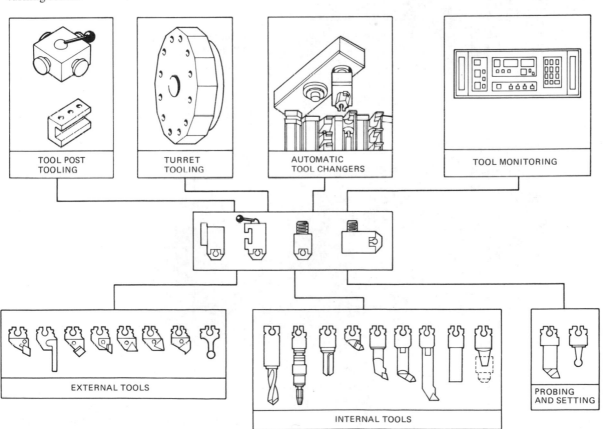

**Tooling system for a
CNC machining centre**
*(Courtesy: Sandvik UK
Ltd)*

**Tooling system for a
CNC turning centre**
*(Courtesy: Sandvik UK
Ltd)*

6.1/7 Automatic tool magazines

Many *CNC machining centres* are provided with **automatic tool changing** (ATC) facilities. This means that tools can be selected and changed under the control of the part program. This facility leads to increased productivity and provides the potential for unmanned machining operation.

The tools, previously identified for the job by the part programmer, must be provided in an automatic tool magazine. An **automatic tool magazine** is an indexable storage facility integral with the machine tool. The prime purpose is to store tools not currently in use but which may be needed for further machining operations or to store "sister" tools for replacing worn or broken tools as these are detected.

When a new tool is commanded, from the part program, the following sequence of events typically take place:

a) The spindle is withdrawn.
b) The spindle or slide moves to a pre-determined tool change position.
c) The spindle (and coolant) is turned off.
d) The ATC causes the required tool to be indexed to the tool change position (often by the quickest route).
e) The current tool is released from the spindle and a mechanical arm proceeds to swap it with the requested tool.
f) The spindle (and coolant) are switched on and machining recommences.

Tools can be identified automatically by coding rings provided on the toolholder itself. These coding rings can be re-positioned to form new codes. Alternatively a coded key (in the form of a machined or ribbed bar) can accompany (and identify) the tool in the tool magazine. These systems allow the tools to be placed at random in the tool magazine. The disadvantage of this arrangement is that a tool change, calling for a new tool, requires the tool magazine to be searched tool by tool, and the coding device decoded. This can be unnecessarily time-consuming because of excessive magazine indexing.

More commonly, the tools have to be placed in the correctly numbered magazine position. A tool change calling tool number 5 really means: change the current tool for the tool that is in magazine position number 5. Care must obviously be exercised when initially loading or subsequently replenishing the tool magazine to ensure that the correct tools are placed in the correctly numbered location. It is the responsibility of the part programmer to give each tool used a unique identity, in the form of a tool number (TØ1, TØ3, etc.), which will be referenced by the part program during machining.

Tool magazines can accommodate between 20 and 160 tools depending on the machine design. The magazines themselves can take different forms. Common configurations include rotary carousels, chain, drum and "egg-box" type magazines. These are illustrated in Fig. 6/13.

Coolant is normally supplied by external supply tubes. These may be designed as a single pipe supplying a jet of coolant directed at the point required. Alternatively, a circular tube, surrounding the cutter and having small holes drilled around the periphery, can provide coolant supply to the area around the cutter. Flood or mist (fine spray) coolant application is usually available, selectable under program control.

Fig. 6/13 Automatic
tool magazines

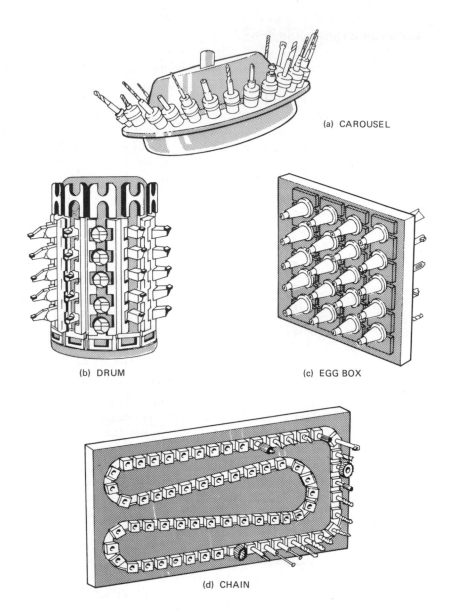

(a) CAROUSEL

(b) DRUM

(c) EGG BOX

(d) CHAIN

In some machining installations, designed for unmanned operation, more cutting tools are often required than can be accommodated in the tool magazine. In such cases the complete magazine can sometimes be detached from the machine and replaced by a fully replenished magazine. In an unmanned environment this task could be accomplished by a robot.

CNC turning centres normally utilise an indexable tool turret. Cutting tools are not exchanged, they are merely indexed to the cutting position. Care must be taken to ensure that the correctly numbered tool (as specified by the part programmer) is located at the correct tool station on the turret. Like machining centres, the slide will be retracted to a pre-determined turret index position prior to indexing taking place. This strategy has to be adopted for fear of crashing the rotating tools into the workpiece or machine structure.

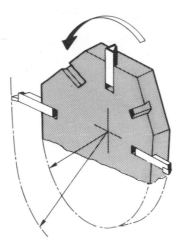 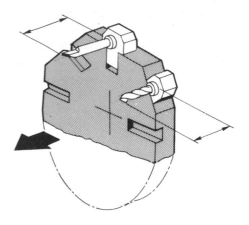

Fig. 6/14a Effective indexing radius of turret increased by external turning tools

Fig. 6/14b Effective slide movement possibly restricted by turning tools that protrude forward of the turret

Design of the turrets allows coolant to be supplied to any tool station currently positioned as the machining station. Pressurised flood or mist coolant can normally be selected from within the part program.

Turning tools located in the tool turret usually project out radially from the turret. They have the effect of increasing the effective turret diameter as far as indexing is concerned. This is one reason for retracting to a safe position prior to indexing. Drilling or boring tools located in the same turret project out at right angles to the turret. The amount of this projection depends on the length of the drill or boring tool employed. This may limit the effective forward movement of the turret. The danger is not of the turret itself colliding with the rotating chuck or workpiece, but of the projecting drills or boring bars doing so.

The situation is made more acute by the fact that the tools most likely to be involved in such a collision are those located in dormant (non-active) turret positions. The programmers (and operators) attention may be focused on the cutting tool currently carrying out a machining operation. It is easy to forget about tools (especially those that project forward) in other turret positions. The effect of mounted tools is shown in Fig. 6/14.

Extreme care must be taken when turning tools and drilling, boring, reaming or tapping tools are employed within the same turret set-up. The programmer and the operator should be especially vigilant of the tool path movements of tools that project forward of the tool turret. There is a risk of collision between the rotating chuck/workpiece and projecting tools in dormant tool stations.

6.2 Speeds and feeds

6.2/0 Choice of speeds and feeds

Some of the earliest decisions a part programmer has to make when planning machining sequences concern the choice of machine speeds and feeds. The choice of correct speed and feed values depends on many factors not least of which is the experience accumulated, over many years, by skilled operators and part programmers. In the final event, both programmed speed and feed values will need modification at the machine to suit local cutting conditions. The purpose of this section is twofold. Firstly, to discuss those factors that influence the choice of speed and feed values. Secondly, to give enough knowledge to access and interpret machining data in order to determine sensible speed and feed values for various machining operations.

The **cutting speed** for a particular machining operation refers to the speed at which the cutting edge of the tool passes over the surface of the workpiece. It may also be referred to as the *surface speed*. It is normally quoted in metres per minute (m/min). The following factors influence the cutting speed:

a) Workpiece material
Hard, strong materials require a lower cutting speed than soft and ductile materials. Reasons for this concern the power capabilities of the machine tool, the strength and wear resistance of the tool, and the cutting action exhibited by the material.

b) Tool material
Special cutting tool materials permit higher surface speeds than conventional carbon or high speed steels. This is because such materials retain their hardness, and other cutting characteristics (such as wear resistance), at the elevated temperatures associated with high cutting speeds.

c) Cutting fluid
The use of the correct cutting fluid can improve cutting conditions. Cooling, lubrication, resistance to chip welding and assistance with swarf removal are all properties exhibited by the correct cutting fluid that combine to allow optimum feeds and speeds to be achieved.

d) Condition of the machine tool
An old machine tool with many worn or loose parts, or an inadequately supported cutting tool or workpiece, will not be able to sustain high cutting speeds. Both these conditions are prone to vibrations set up during machining. Inferior surface finish, poor dimensional accuracy and excessive tool wear would prevent optimum feeds and speeds from being used.

e) Volume of material to be removed
In general, a light finishing cut with a fine feed can be run at higher speeds than a heavy roughing cut. Surface finish requirements, tool life and speed of production determine what depths of cut and feeds can be permitted.

f) Power available at the spindle
It is possible by employing modern cutting tool materials and optimistic choice of feed and speed values to demand power requirements for cutting that exceed the power capability of the machine tool. There is a strong argument for calculating machining requirements first and using these as the basis for specifying the size of the machine tool required.

Before choice of the correct speeds and feeds can be made, the term "correct" must be defined. A number of conflicting criteria may suggest a number of "correct" values. It depends which of the following factors are to be given maximum priority:

1) To minimise production cycle time.
2) To maximise tool life between tool changes.
3) To maintain close dimensional accuracy.
4) To maintain high-quality surface finish.
5) To maximise material removal rate.
6) To minimise tool breakages.
7) To sustain long periods of uninterrupted operation.
8) To minimise wear and stress on the machine tool.
9) To minimise machining cost per piece.

The above factors together with point *d* above are the reasons why the part programmer's choice of speed and feed values will be modified at the machine tool by the operator.

6.2/1 Cutting speed and spindle speed calculations

Cutting tool manufacturers and suppliers will supply, on request, comprehensive machining data for their products. This should be the starting point for determining machine speeds and feeds. Since the manufacturer can only be sure of points *a*, *b* and *e* above, the machining data supplied represents the best guide on which to base the final choice of values. It will usually be stated that the machining data supplied is "under 'optimum' cutting conditions" or "for 'average' machining conditions". Adjustment of the stated values may then be made, either up or down, in the light of the factors discussed previously.

For a particular workpiece-material/tool-material combination, a cutting speed in m/min will usually be quoted. This value will have to be converted into a spindle speed for the machine and machining operation concerned. The **cutting speed** is the relative surface speed at which the cutting edge passes over the surface of the workpiece. It does not matter whether the tool rotates, as in milling or drilling operations, or the workpiece rotates, as in turning or boring operations. The following discussion relates to a rotating workpiece. For calculating spindle speeds for milling or drilling operations, "cutter" can be substituted for "workpiece".

The actual **spindle speed** to be set, which will maintain the quoted surface speed, depends on the diameter of the workpiece. A large diameter means a large circumference, a smaller diameter means a smaller circumference of the workpiece. If both a large and a small diameter workpiece are to be turned at the same surface speed, a moments thought will confirm that the smaller diameter must rotate more quickly.

A standard formula relates the surface cutting speed with spindle speed:

Spindle speed (rev/min)

$$= \frac{1000 \times \text{Surface cutting speed (m/min)}}{\pi \times \text{Workpiece or cutter diameter (mm)}}$$

127

Note the following points about the above formula:

(i) The constant 1000 serves to convert the linear units of the workpiece or cutter diameter, usually quoted in millimetres, into the same linear units of the cutting speed, usually quoted in metres.

(ii) The constant π, multiplied by the workpiece or cutter diameter gives the circumference of the workpiece or cutter.

(iii) The abbreviation for spindle speed should be quoted as rev/min and not as the once familiar RPM.

(iv) Common (although not ISO) abbreviations are S for surface cutting speed, N for spindle speed and d for workpiece/tool diameter.

The formula may then be more generally quoted as:

$$N = \frac{1000 \times S}{\pi \times d}$$

Values of cutting speed or diameter to be used in the above formula may, on occasions, be quoted in different units. If this is the case then appropriate modification to the formula may be required.

As a guide, the following cutting speeds are suitable for "average" machining conditions.

| Cutting Tool Material | Surface cutting speeds (metres/minute) | | | |
| | Workpiece Material | | | |
	A1 Alloy	Brass	Cast Iron	Mild Steel
HSS	120	75	18	30
Carbide	500	180	120	200

To illustrate the application of the above formula, consider the following worked example.

EXAMPLE   Calculate the spindle speed required to turn a 75 mm diameter shoulder on a mild steel component using an HSS tool. What percentage increase in cutting speed can be achieved by using a carbide tool?

$$N = \frac{1000 \times S}{\pi \times d} \qquad \text{state formula}$$

HSS tool

$$N \,(\text{rev/min}) = \frac{1000 \times 30}{\pi \times 75} \qquad \text{substitute known values for HSS tool and m/steel workpiece}$$

$$= \frac{30\,000}{235.71} \qquad \text{carefully perform calculations step by step}$$

$$N = 127 \,\text{rev/min} \qquad \text{determine spindle speed for HSS tool}$$

Carbide tool

$$N\,(\text{rev/min}) = \frac{1000 \times 200}{\pi \times 75}$$ substitute known values for carbide tool

$$= \frac{200\,000}{235.71}$$ carefully perform calculations step by step

$$\underline{N = 848\,\text{rev/min}}$$ determine spindle speed for carbide tool

$$\% \text{ increase in spindle speed} = \frac{848}{127} \times 100 = 668\%$$

The calculated spindle speeds will be correct for producing sliding cuts along the longitudinal axis of the workpiece. When performing such cuts, the diameter d remains constant. When performing facing cuts along the end of the workpiece however, this is not the case. The diameter at which cutting is taking place continuously decreases to zero as the tool approaches the centre of the bar. It follows that, as the workpiece diameter decreases, the spindle speed should increase in direct proportion. The above calculated speed will only be valid when turning at the specified diameter (75 mm). Most CNC machine tools have spindle speeds infinitely variable between their upper and lower speed boundaries. In many cases the control unit is capable of monitoring workpiece diameter and continuously adjusting the spindle speed accordingly. This condition can be selected under program control, and is known as *constant surface speed* (CSS) *machining*. This is dealt with more fully in Chapter 7.

6.2/2 Feed rate calculations

The **feed rate** governs how quickly, or how slowly, the cutting tool is traversed across the workpiece surface. The feed and the depth of cut determine the volume of material removed. In general, it is more efficient to remove material with large depths of cut and slower feed rates than vice versa.

In terms of productivity, feed rates should be as high as possible consistent with local machining conditions. In practice an upper limit will be encountered because of the following considerations:

a) The power available to drive the tool/table.
b) Maintaining straight, perpendicular and accurate features due to tool, workpiece or structural deflections.
c) Achieving minimum levels of surface finish.
d) Minimising the risk of tool breakage.
e) Maximising tool life.
f) Limiting vibration due to tooling or machine tool deficiencies.
g) Chipbreaking capacity of insert.

Where either spindle speed or feed rate can be increased, it is more desirable to increase the feed rate. When employing carbides, minimum feed rates should not be less than 0.1 mm/rev.

Feed rates are normally quoted in millimetres per spindle revolution (mm/rev) or millimetres per minute (mm/min). In cases where it is necessary to convert from one set of quoted units to the other, the following formulae may be applied:

Feed rate (mm/min) = Feed rate (mm/rev) × Spindle speed (rev/min)

or

$$\text{Feed rate (mm/rev)} = \frac{\text{Feed rate (mm/min)}}{\text{Spindle speed (rev/min)}}$$

Consider the following worked example.

EXAMPLE   In a turning operation the spindle speed is 200 rev/min. The planning sheet specifies a feed rate of 0.2 mm/rev. Calculate the feed rate in mm/min to be programmed into the machine.

Feed rate (mm/min) = Feed rate (mm/rev) × Spindle speed (rev/min)
= 0.2 × 200

Feed rate = 40 mm/min

When turning under constant surface speed conditions, feed rate values will be applied to the CNC control unit in mm/rev. This will ensure that the feed rate remains constant when the spindle speed automatically increases or decreases according to the diameter being machined. Feed rate values also depend on particular tool/workpiece material combinations. Machining data from manufacturers' literature should be sought for initial values to be specified. Some examples, for guidance only, are specified in Fig. 6/15.

When inserted tooth milling cutters are specified, it is more common to state feed rate values in millimetres/tooth (mm/tooth). This is preferred for the following reason. Consider two milling cutters of the same diameter, but having a different number of teeth. Using a feed rate quoted in mm/rev, mm/min or m/min, the cutter having the smaller number of inserts will remove more material per tooth during cutting. Quoted feed rate values in mm/rev, mm/min or m/min are thus specific to individual cutters and cannot be generally applied. This is an unsatisfactory state of affairs since identical inserts can be used in different cutters. The machining data should thus be transferable irrespective of the number of teeth possessed by the cutter.

Feed rate values quoted in mm/tooth need converting into the units of mm/rev, mm/min or m/min for entry into the CNC control unit. The following simple formulae may be applied:

Feed rate (mm/rev) = Feed rate (mm/tooth) × Number of teeth

or

Feed rate (mm/min) = Feed rate (mm/rev) × Spindle speed (rev/min)

The following worked example illustrates the approach.

EXAMPLE   An inserted tooth face milling cutter has 10 teeth. It is to be used to mill a surface using a spindle speed of 1150 rev/min and a feed rate

Fig. 6/15 Machining data

MILLING	FEED RATE mm/tooth			
	HSS		SINTERED CARBIDE	
WORKPIECE MATERIAL	END MILLS AND SLOT DRILLS	FACE AND SHELL END MILLS	END MILLS AND SLOT DRILLS	FACE AND SHELL END MILLS
MILD STEEL	0.13	0.25	0.25	0.50
CAST IRON	0.20	0.40	0.25	0.50
BRASS	0.18	0.36	0.15	0.30
AL. ALLOY	0.28	0.56	0.25	0.50

TURNING	FEED RATE mm/rev	
WORKPIECE MATERIAL	HSS	SINTERED CARBIDE
MILD STEEL	0.20	0.80
CAST IRON	0.40	1.00
BRASS	0.80	1.50
AL. ALLOY	0.30	1.00

DRILLING	FEED RATE mm/rev	
DRILL DIA. (mm)	HSS	SINTERED CARBIDE
2	0.05	0.15
4	0.10	0.15
6	0.12	0.15
8	0.15	0.15
10	0.18	0.25
12	0.20	0.25
14	0.22	0.25
16	0.25	0.32
18	0.28	0.32
20	0.30	0.32

of 0.2 mm/tooth. Calculate the feed rate in m/min that should be programmed.

Feed rate (mm/rev) = Feed rate (mm/tooth) × Number of teeth

$$= 0.2 \times 10$$

Feed rate = 2 mm/rev

Feed rate (mm/min) = Feed rate (mm/rev) × Spindle speed (rev/min)

$$= 2 \times 1150$$

$$= 2300 \, \text{mm/min}$$

Feed rate (m/min) $= \dfrac{\text{Feed rate (mm/min)}}{1000}$

Feed rate – 2.3 m/min to be programmed

Whenever particular speed/feed combinations have been determined, the power requirement should be calculated and then verified that it is towards the upper end of the machine tool's capacity. If it is far short then higher rates can be specified since machining capacity is being wasted.

6.2/3 Feed rate modification for contours

The programmed feed rate relates to the feed at which the centre of the cutter traverses the cutter path. When contour milling circular contours, the programmed feed rate will need modification. The reason is as follows.

Consider the case where an external circular contour has to be machined. Let us assume that the "part radius" to be machined is 60 mm and that we are using a 30 mm diameter end mill. The centre of the cutter will be traversing along a radius of 75 mm, yet the part radius required is only 60 mm. The feed rate is effectively being used to machine a larger component. In the case of external circular contours, the feed rate will need to be increased. Conversely, when machining internal circles, the feed rate will need to be decreased.

The amount of modification depends on the ratio of the part radius and the radius traversed by the cutter path. This will be influenced by the cutter diameter. The following formulae may be applied:

$$\text{Increase feed rate by} \quad \times \frac{(PR + CR)}{PR} \quad \text{for } external \text{ circles}$$

$$\text{Decrease feed rate by} \quad \times \frac{(PR - CR)}{PR} \quad \text{for } internal \text{ circles}$$

where PR = Part Radius
CR = Cutter Radius

EXAMPLE   An external circular contour of radius 120 mm is to be machined using a 30 mm diameter end mill. Before reaching this point in the part program the programmed feed rate is 300 mm/min. Calculate the new feed rate to be programmed.

External circles, so feed rate must be INCREASED.

$$\text{Increase in feed rate} = \frac{(PR + CR)}{PR}$$

$$= \frac{120 + 15}{120}$$

Increase = 1.125

New feed rate = Old feed rate × Increase
$$= 300 \times 1.125 \text{ mm/min}$$

New feed rate = 338 mm/min

6.2/4 Tool life calculations

Tool life is important for the following reasons:

a) It can influence the choice of speed and feed values.
b) It is important for costing purposes.

c) It indicates tool change periods to maintain accuracy and the quality of surface finish on machined components.
d) It assists in planned tool replacement to ensure smooth production.
e) It assists in the planning and maintenance of tool stocks.
f) It may assist in reducing scrap, or damaged components, due to catastrophic tool failure.

Manufacturers' machining data for speeds and feeds is usually quoted assuming a specified tool life. Tool life is normally quoted in minutes. Production engineers may extrapolate this information throughout a machining cycle, and produce tool life figures in terms of number of components machined. A typical tool life for a carbide insert may state 15 minutes per cutting edge. Whilst this may appear to be short, this is the time that the cutting edge is actually in contact with the workpiece and not the amount of time it is resident in the machine tool.

Here is an interesting, simple exercise to carry out in this respect. Firstly, with a stop watch, time how long it takes to produce a machined component on, say, a turning centre. Repeat the timing but this time only record the time that the tool is in contact with the workpiece. The results may confirm that 15 minutes is a reasonable life for a cutting edge. When using hardmetal inserts this time can be multiplied by the number of cutting edges present on the insert.

The determination of tool life is a difficult problem. A decision has to be taken as to what constitutes the end of the useful life of a cutting tool. Generally, this is signalled by the onset of a particular condition. For example:

a) The tool fails by catastrophic failure due to the combined effects of force and temperature.
b) Machined surface texture becomes unacceptable.
c) Dimensional tolerances cannot be maintained.
d) Cutting forces may increase and cause unacceptable deflections or vibrations.
e) The dimensions of the tool tip change by a significant amount.

The last two points are important in the field of adaptive control techniques. (Adaptive control is discussed in Chapter 8, section 8.3.)

Tool life may be predicted using the classical *tool life equation* as follows:

$$VT^n = C$$

where V = cutting speed (m/min)
T = tool life (min)
n = an index closely related to the cutting tool material
e.g. 0.1 to 0.15 for HSS
0.2 to 0.4 for tungsten carbide tools
0.4 to 0.6 for ceramic tools
C = a constant based on machining tests with fixed feed rate, depth of cut, rake angle, tool geometry, workpiece material, etc.

It must be stressed that this formula is empirical (based on experiments) and can only be used as a guide.

The values of *n* and *C* are determined by cutting tests for particular machining condition on particular materials. The values will not be valid if any of the machining conditions (such as feed or depth of cut) are changed.

By way of illustrating the technique, consider the following worked example.

EXAMPLE   Calculate the life of a tungsten carbide cutting tool when machining a cast iron component. The cutting speed used is 80 m/min and values for n and C are 0.25 and 150 respectively.

$$VT^n = C \quad \text{so that } T^n = \frac{C}{V}$$

Therefore

$$T = \sqrt[n]{\frac{C}{V}}$$

$$= \sqrt[0.25]{\frac{150}{80}}$$

$$= \sqrt[0.25]{1.875}$$

Evaluate root by logarithms:

  a) Find log of 1.875 in log tables.
  b) Divide log of 1.875 by 0.25
  c) Find antilog of result.

$$\log 1.875 = 0.2729$$

$$\frac{0.2729}{0.25} = 1.0196$$

$$\text{Antilog } 1.0196 = 12.35$$

$$\underline{\text{Tool Life } T = 12.35 \text{ min } (12 \text{ min } 21 \text{ sec})}$$

6.3 Workholding and setting

6.3/0 Workholding for CNC

Efficient **workholding** is central to any successful machining operation if accurate, consistent components are to be produced at minimum cost. With CNC machining techniques, extra demands are made on traditional methods of workholding for the following reasons.

1 Many CNC machines are capable of performing many different operations, using many different cutting tools, on many different faces of a component, at one setting. This means that access to all sides of a component becomes necessary, preferably without the need to perform time-consuming clamp changes or component re-positioning.

2 Multi-directional cutting forces exerted on the workholding device will demand greater rigidity and strength in all directions.

3 Automated component handling is becoming more widespread in CNC installations. Workholding devices have need to be adaptable and lend themselves to loading and unloading by robot devices.

4 Flexible manufacturing concepts rely on quick changeover times, from job to job, at short notice. An ideal workholding device should be capable of accommodating a number of different components without elaborate modification.

The ideal workholding device, for CNC applications, should embody a number of desirable features:

a) Provide positive location of the component such that it cannot move or rotate.

b) Support and secure the component such that it will not deflect, distort or be marked due to the application of clamping or cutting forces.

c) Provide accurate, repeatable location and ensure that the component cannot be incorrectly loaded.

d) Be quick and simple to operate, to reduce handling time.

e) Not interfere with the operation or movements of the machine tool or its associated guarding.

f) Be cheap and adaptable to a number of different applications.

g) Be designed to permit a number of machining operations to be carried out, in a number of planes, at one setting.

h) Enable a number of components to be machined simultaneously.

i) Be adaptable to automated component handling.

j) Provide for easy cleaning and removal of swarf, preferably without human intervention.

k) BE SAFE! Especially when pneumatic, hydraulic, magnetic or electrical actuation is employed.

The aspiring production engineer will quickly realise that a workholding device embracing all the above qualities has yet to be invented. Compromise is inevitable. The aim should be to maximise the benefits and minimise the limitations of the above features, within the objectives and priorities sought for the manufacturing operations concerned.

6.3/1 Principles of location

The first objective of any workholding device is that it should consistently locate the component accurately and positively.

Any body in space has **six degrees of freedom**. It can have linear movement along 3 axes and rotational movement about the same 3 axes. This is illustrated in Fig. 6/16a. The first, and most important, principle of location is that the six degrees of freedom should be reduced to zero. Fig. 6/16b illustrates that the six degrees of freedom can be reduced to zero by employing, at maximum, a six-point location principle.

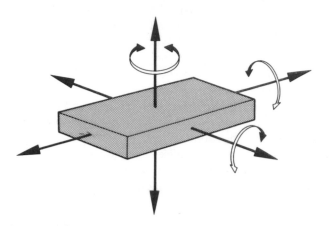

Fig. 6/16a Six degrees of freedom

Fig. 6/16b Principle of six-point location

The diagram assumes that gravity is holding the component into the locations. In practice a clamping device would be employed. If more than six points are provided, the additional points will be redundant. The use of three location points in the horizontal plane ensures that an uneven component (such as a casting or a forging) will seat without rocking. In practice, six locations may not actually be required but the principle remains the same. Locating features may employ positive locations, in the form of physical restraints, or may rely on the action of friction in conjunction with a clamping device.

When designing or evaluating location features for workholding devices, the following points should be considered:

a) Design locations to reduce the six degrees of freedom to zero with no redundant locations.
b) Locate from the same machined surface (datum) for as many operations as possible to reduce the possibility of error.
c) With first-operation work on rough unmachined surfaces, use three-point location where possible. Adjustable or expanding locators may be used to accommodate large variations in size.
d) Ensure the location is "foolproof" such that the component can only be located in the correct position.
e) Location features should not be swarf traps and they should have clearance provided, where necessary, to clear machining burrs.

f) Consider the manipulative difficulties in loading or unloading the component. Retractable or removable features may be necessary. Flatted peg locations enable bores to be located and removed more easily.

g) All location features should be safe to the operator. Burrs, sharp corners, finger traps and other potential hazards should be eliminated.

h) Locating features should be hardened and ground to ensure consistent accuracy and help minimise wear.

Fig. 6/17 A simple nesting block assists with component location

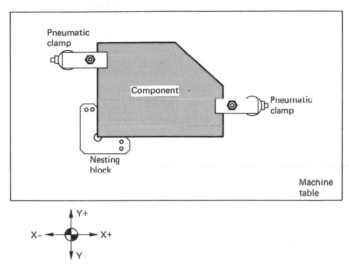

Simple dowels or *nesting blocks* often provide sufficient location for most components. The use of *grid plates* to assist in component location is discussed in section 6.3/5. A simple nesting block is illustrated in Fig. 6/17.

6.3/2 Principles of clamping

Clamping of components is provided to restrain the component from moving during machining operations. Forces encountered include the weight of the component and workholding device itself, applied forces due to the machining operation, and inertial forces caused by spindle rotation or slide movement.

Clamping is important since it can influence:

The accuracy of the component produced.
The quality of the component produced.
The speed of the operation being carried out.
The life of the cutting tools.
The safety of the operator.

Clamping arrangements must avoid distortion to the workpiece, the workholding device, the location features used to guarantee accuracy of the machining operation, and the clamps themselves. For this reason clamping forces should not be excessive but sufficient to hold the work rigidly. They must be applied to a solid or supported part of the workpiece. For CNC applications, automatic operation is desirable. This not only ensures quick operation but eliminates the tendency to overtighten nuts and bolts.

When designing or evaluating clamping features for workholding devices, the following points should be considered:

a) Clamps should be applied to the component where it is rigid and well supported. Extra support should be provided where necessary.

b) Clamping forces should be controlled such that they do not distort the workholding-device/workpiece/clamping-device but are sufficient to resist cutting forces.

c) Avoid cutting against the clamps.

d) Ensure that clamps do not unduly impede likely cutter paths, and that sufficient clearance is available (when they are released) to load and unload the workpiece safely. Consider the use of ejectors.

e) Clamps should be quick-acting consistent with considerations of cost. Power-operated devices (pneumatic/hydraulic) should be fail-safe. Mechanical devices should show due regard to safety, ease of use and operator fatigue.

f) When clamping through a bore, position stops to prevent the workpiece from rotating under the action of the applied cutting forces.

g) If possible, use more than one clamp.

h) Clamping bars supported by springs will not drop when the component is removed.

6.3/3 Workholding devices

Workholding devices can be broadly divided into two groups: those used for rotating workpieces, and those used for fixed workpieces on machines that employ rotating cutters.

Consider the workholding devices used for ROTATING WORKPIECES.

CHUCKS **Chucks** may be of the three-jaw or four-jaw variety. Three-jaw chucks are used for general-purpose applications, on round workpieces, largely because of their self-centring ability. Four-jaw chucks are used where asymmetrical or non-round workpieces need to be machined, or where superior gripping power is required. Each jaw of a four-jaw chuck has to be set independently. This is a time-consuming operation and requires a skilled setter. For these reasons they are not often employed for CNC work. If irregular-shaped work is to be turned, it is more likely that a special-purpose turning fixture will be designed.

Chucks may be manually operated or power assisted. It is more common for power-assisted chucks to be used for CNC work where the emphasis is on speed of loading and unloading. Power assistance may be by pneumatic or hydraulic operation, the latter being used for larger applications. Greater gripping power is obtained using hydraulics since the hydraulic fluid is essentially incompressible. Such chucks have a jaw movement of only a few millimetres and so must be initially set for the diameter of workpiece being machined. Automatic chucks are normally operated by a foot pedal but a "chuck enable" button, on the operating console, has to be depressed before the chuck can be released. This is a safety device to prevent accidental misoperation of the chuck. Bar feeders may be employed where many identical components are required, and the raw material can be obtained in bar stock form.

When loading components into chucks, it is desirable to locate the component against the back face of the chuck, or a suitably designed spacer. This ensures positive location to resist the applied cutting forces. Where the component is not backed up in this way, the operation is relying on the frictional location provided by the chuck jaws alone.

COLLETS   **Collets** or collet chucks are quick-acting fixed-diameter work-holding devices. They are designed for holding close diameter round components. If components are to be machined from lengths of bright bar, the bar stock can be fed through the centre of the collet onto a fixed stop. Collets offer quick, positive and constant re-chucking and afford a wide area of contact for gripping. Because they are of fixed diameter, a set is required to accommodate different-diameter workpieces. Collets may have jaws of different forms to accommodate different sections of component.

Fig. 6/18 Collet details

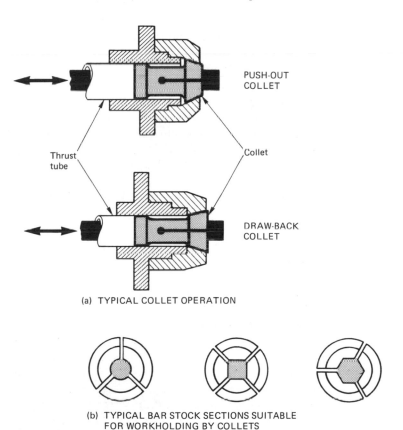

(a) TYPICAL COLLET OPERATION

(b) TYPICAL BAR STOCK SECTIONS SUITABLE
     FOR WORKHOLDING BY COLLETS

Collets operate on the principle of movement along a taper. There are different designs in that the taper may be *pushed* (as in a push-out collet), or *pulled* (as in a draw-back collet). They may be manually operated or power assisted.

Bar feeders represent a useful addition to a collet set up for speed of component feeding.

Collet details are illustrated in Fig. 6/18.

CENTRES and FACE DRIVERS Where longer components are being machined then support can be provided by a rotating (live) **centre** held in a tailstock. It is likely that only the larger turning centres will have tailstocks. One of the major functions of the tailstock on conventional machines is that of holding drills, taps and reamers for producing holes. On CNC machine tools these operations can be accommodated from within the turret. For this reason many of the smaller CNC machines no longer have tailstocks.

In conjunction with the centre, for long and/or slender components, steadies can also support the workpiece during machining. These are employed in much the same way as they would be on conventional lathes.

When employing a centre as a support, at the tailstock end of the workpiece, the workpiece has to be driven at the headstock end. In some cases the workpiece is simply held in, and driven by a conventional chuck. This is acceptable when machining does not have to be carried out along the entire length of the workpiece. Where this is not the case, and a chuck cannot be used, alternative means of driving the workpiece have to be considered. A common solution is to use a **face driver**. This is a device, mounted in the spindle, which transmits the rotation of the spindle to the workpiece via radial driving pins protruding from the face of the driver. An axial force, created by the travel of the tailstock causes the workpiece to be pushed (and gripped) against the face driver. The driving pins are backed up by a closed circuit of hydraulic fluid which allows movement of the pins to accommodate slight irregularities in the front face of the workpiece. In this way equalisation of pressure ensures that all driving pins come into contact with the workpiece. Some face drivers can be obtained with a polymer-type backing medium rather than a closed hydraulic circuit.

The principle of the face driver is illustrated in Fig. 6/19.

Fig. 6/19 Principle of the face driver

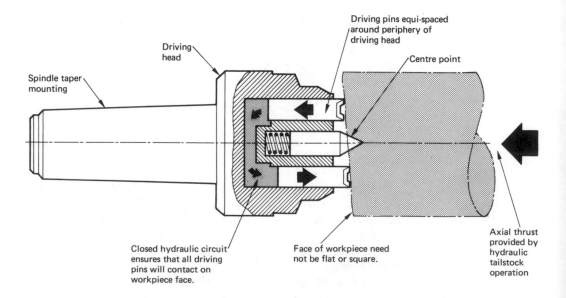

Driving pins equi-spaced around periphery of driving head

Driving head

Centre point

Spindle taper mounting

Closed hydraulic circuit ensures that all driving pins will contact on workpiece face.

Face of workpiece need not be flat or square.

Axial thrust provided by hydraulic tailstock operation

MANDRELS Fixed **mandrels**, working on the principle of the taper, can be used to clamp the workpiece on its inside diameter. This allows the entire length of the workpiece to be machined but means that it must be provided with an accurate bore. Clamping on a fixed mandrel does not permit accurate axial positioning of the workpiece and a press must be used to mount the workpiece.

Expanding mandrels may offer a more acceptable alternative where accurate location of the workpiece is required.

Mandrels can be spindle mounted (using a suitable adapter), flange mounted, or driven between centres via a catchplate and driving dog arrangement.

Face drivers, mandrels and centres
(Courtesy: Sandvik UK Ltd)

TURNING FIXTURES   Where the component is unusually large or irregular (as in the case of castings or forgings), special-purpose **turning fixtures** may have to be designed. These will be designed embodying the principles of location and clamping discussed previously. The fixtures will then be mounted directly onto the spindle of the machine itself, or on a faceplate mounted on the spindle. It would be unusual to employ faceplates alone on CNC machines since the setting-up time would be prohibitively long. Robotic devices can easily cope with loading and unloading purpose-designed fixtures employing power operated clamping devices.

Workholding devices used on machine tools employing ROTATING CUTTERS can be varied. In all cases the component remains essentially fixed. A number of general considerations apply to all methods of workholding.

In all cases the workholding device should be positioned at the centre of the worktable in the X-axis on machining centres. This ensures the greatest support and minimises any static deflection of the machine table due to the weight of the component and the workholding device. Similarly, positioning the workholding device as near to the column of the machine as practicable minimises any deflection of the machine tool structure due to the effects of overhang. The direction of the cutting forces should always be directed towards a positive, fixed location. Dowels are useful location devices in this respect since they offer positive location in all directions. Finally, if there is flexibility to position the workpiece or the table, other factors that might be considered concern the ease of loading and unloading and the ease of swarf removal from the cutting zone. The component should be positioned such that the cutting action directs the machined swarf away from the operator.

The common workholding methods are discussed below.

MACHINE VICE   Perhaps the most versatile workholding device for small prismatic components is the familiar **machine vice.** This offers simplicity, versatility, rigidity; it can easily be adapted to power-assisted operation and it is readily available in a range of sizes at a reasonable cost. Extra flexibility may be offered by the use of a swivel vice (allowing rotation in the horizontal plane), or a universal vice (allowing rotation in both horizontal and vertical planes). The vice should be clamped to the worktable in such a way that the component is positively located (to resist cutting forces) against the fixed jaw of the vice. Relying on the frictional location of the vice jaws is not recommended.

Specially machined replaceable vice jaws can enhance the location and clamping ability of the machine vice, at minimal cost.

CLAMPING SET   **Clamping sets** comprise a range of modular components which can be assembled to form workholding devices. A minimum set would include a range of studs, nuts, washers, clamping strips, packing pieces or step blocks, tee-nuts/bolts, etc. They are normally used in conjunction with other standard items of workshop equipment for supporting the work—for example, parallel bars, vee-blocks, angle plates, screw jacks, and so on.

The most common set-ups are those of the "bridge" or "strap" clamping arrangement or an "edge" clamping arrangement. The latter is preferred where it is required to machine the whole of the top face of a component

Fig. 6/20 Clamping set applications

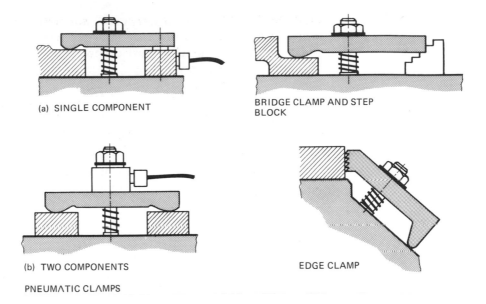

(a) SINGLE COMPONENT

BRIDGE CLAMP AND STEP BLOCK

(b) TWO COMPONENTS

PNEUMATIC CLAMPS

EDGE CLAMP

without necessitating an intermediate clamp change. These are illustrated in Fig. 6/20. When used in these configurations, certain points should be observed:

a) Position studs/bolts as close as possible to the workpiece.

b) Pack the rear of the clamp until it is level with, or slightly higher than, the height of the workpiece—never lower!

c) Position clamps so that the stud is closer to the workpiece than it is to the packing block.

d) Select studs that are as short as possible but long enough for the nut to be fully engaged on the thread.

e) Always use spherical clamping washers underneath the clamping nuts.

f) Ensure that all clamps and packing clear, and do not impede, the intended cutter path.

g) Always clamp on a solid part of the workpiece; use supporting devices where necessary.

h) Use more than one clamp where possible.

i) Before moving clamps, after partial machining, ensure that one or more other clamps are still holding the workpiece in position.

j) Springs should be inserted between the machine table and the strap clamp to support the clamp during loading and unloading.

Clamping sets are primarily workholding devices. The component usually has to be positioned and/or located by additional means. Examples of such means are discussed in section 6.3/5. The selection and usage of clamping sets is largely a manual operation. It is possible to automate the operation by using small pneumatic cylinders instead of clamping nuts. They find greatest application in CNC work where components can be loaded away from the machine on *sub-tables* or *pallets*. Ready-loaded pallets can then be exchanged rapidly at the machine tool. (Automatic pallet changing (APC) devices are discussed in Chapter 8, under Flexible Manufacturing Systems.)

FIXTURES   Special-purpose fixtures embodying the principles of work location and clamping discussed earlier, are often employed. The decision to design and manufacture such a fixture will depend on such factors as:

a) The size, shape and form of the component or raw material.
b) Suitability, or otherwise, for efficient workholding (and setting) by other means.
c) The number of components required and the likelihood of repeat orders.
d) Anticipated increase in productivity.
e) Projected cost of producing and maintaining the fixture.
f) Lead time for components and/or to design and manufacture the fixture.
g) The need to coordinate workholding with automated loading and unloading of components.
h) The possibility of machining a number of components at the same set-up.
i)   The need to provide extra degrees of movement not provided by the machine tool itself. For example, rotary indexing of components.

If a fixture is to be specially designed, it is a good idea to incorporate some means for establishing the X∅, Y∅, Z∅ datum position for setting the cutting tool. This could take the form of a simple hardened and ground setting-block integral with the fixture but sufficiently discrete so as not to interfere with the machining operations. The operator would then "touch" on to the setting-block and set the appropriate axis registers to zero. Similar considerations apply to providing locating features for the fixture itself. This is important if the fixture itself is to be located and mounted on a grid plate.

Fixtures may also be built up using standard modular items such as locators, clamps, baseplates, etc. After use, such fixtures can be cannibalised and the components used again.

6.3/4   Other workholding devices

Many advanced CNC machining centres are equipped with the capability to perform simultaneous machining operations in more than three axes. Extra axes of motion may include rotary motion about the primary linear axes X, Y and Z. Where such capability exists, extreme flexibility for the production of complex components is provided. Where it does not exist, it may be necessary to fall back on traditional workholding devices to provide the extra axes of movement. In this category are the dividing head, rotary table, adjustable angle plate, and so on.

In general, such devices are not wholly compatible with the concept of CNC machining techniques. The specific application will determine their use. Where many components require the treatment afforded by such devices, it is often more expedient, and more cost effective, to commission purpose-designed fixtures with built-in actuators.

6.3/5 Setting on CNC machine tools

CNC machines commence machining using positional coordinates stated within the part program. The datum point, about which the dimensions are referenced, must be set within the CNC control unit before machining begins.

Confusion can arise since the term "datum" can apply equally well to a number of different datums. For example, the component may have a datum, the detail drawing may have a datum (which may not be the same as the component datum), the CNC machine tool may have a fixed datum which cannot be moved, or the machine may have no datum unless specifically set by the operator. It is the responsibility of the part programmer to reconcile all these factors to a common datum point that can be set at the machine tool.

In the context of CNC machine setting, the term **datum** refers to the assumed zero positions of the axis movements. It will in fact refer to the datums of all the axis movements. There will be three in the case of a typical machining centre ($X\emptyset$, $Y\emptyset$ and $Z\emptyset$), two in the case of a typical turning centre ($X\emptyset$ and $Z\emptyset$).

The part programmer must therefore ensure that the machine operator receives full documentation on where the component must be positioned, and where the machine datum must be set to, before machining commences. In many cases the part programmer can provide a programmed *dwell* within the part program to accomplish datum setting. This allows axis datum setting to be accomplished and axis position values to be temporarily added to the part program. (This is dealt with in Chapter 7.)

Setting the machine tool for a machining operation involves two considerations. The first concerns positioning and locating the component (or workholding device) on the machine, and the second concerns establishing the machine datum. In the majority of cases, establishing the machine datum is done after the component has been positioned and located on the machine.

1 Positioning the component

On turning centres when standard workholding devices are employed, the components are inherently positioned on the centre-line of the spindle. The centre-line of the spindle is almost universally accepted as being the datum $X\emptyset$ in the X-axis. The datum position in the Z-axis will depend on the particular machine. It may be fixed or fully floating. A common datum position is usually at the back face of the chuck. This is designated $Z\emptyset$ initially since it is a fixed point on the machine tool that will never change. For setting/machining purposes the datum may be temporarily "shifted" to a more convenient position for programming. If a zero shift facility is available then the position of the component, in the Z-axis, is not critical.

On machining centres with fixed datums, the positioning of components becomes somewhat critical. Part programmed dimensions refer to distances from a datum point fixed on the machine structure, rather than a point fixed on the component. It is essential therefore that the component is positioned accurately with respect to the fixed machine datum. A number of methods can ensure this.

A **sub-plate** comprising nesting blocks, or a fixture, can be accurately positioned and fixed on the machine table. All components can then be loaded into the nesting blocks or fixture quickly, and with comparative ease. This enables subsequent components to be positioned repeatedly without the need to set them individually.

For larger components, or components where the expense of a fixture is not justified, a setting aid called a **grid plate** can be used. A grid plate is a base plate in which accurately positioned dowel holes, and tapped fixing holes, are provided. The holes are positioned in a grid or matrix formation at regular intervals, hence the name. Each hole can be referenced by a particular column/row position within the matrix. Column positions might be designated alphabetically (A, B, C, etc.), and rows numerically (1, 2, 3, etc.). Any hole can thus be identified by an unambiguous two-character code. Alternative designs use accurately machined slots, instead of dowel holes, to provide accurate location. The grid plate will, of course, have to be accurately positioned on the machine tool relative to the fixed datum position of the machine. Once set it can be securely clamped. The dowel holes can provide accurate positional locations suitable for use by a variety of setting devices. Fixtures, nesting blocks, stops and components need only be provided with the correct-size dowels, or tenons, to ensure accurate location. Clamping may be accomplished using the tapped holes provided in the grid plate. Grid plates may also be set in the vertical plane if desired. Typical grid plates are illustrated in Fig. 6/21.

2 Setting axis datums

On CNC machines employing fully floating datum facilities the situation is much simpler. It is common to position the workpiece or workholding device on the machine table, for convenience. The tool, or setting probe, is then jogged manually to touch onto the component in each axis in turn. With the tool or setting probe in the correct setting position (component datum position), one of two actions may be taken. The choice will depend on the machine tool.

The first action involves setting the relevant axis register to zero, by entering at the console. A button marked "axis zero" may also have to be depressed to confirm the action. Thereafter, the machine zero position is assumed to be that point. A programmed dwell within the part program may be provided for this setting to be carried out by the operator. A special G-code (called Pre-set Absolute Register) is provided for this purpose. All subsequently programmed positional moves will be made with reference to this zero point.

The second action is perhaps more prevalent on CNC turning centres. At start-up, many turning centres send their slides to a known position normally at the extremes of their axis movements. When at this position, indicated by limit switches detecting the limits of travel, the control unit "knows" that the slides are at known dimensions from the machine datum point. This may be, for example, 800 mm from the back face of the chuck in the Z-axis, and 500 mm from the centre-line of the spindle in the X-axis. The exact dimensions will, of course, vary from machine to machine. The tool can then be jogged to touch on the component as previously described. The dimensions showing on the axis read-outs are then subtracted from the known "limits of travel" dimensions. The resulting values are then entered into axis offset registers

Fig. 6/21a Grid plate using dowel holes

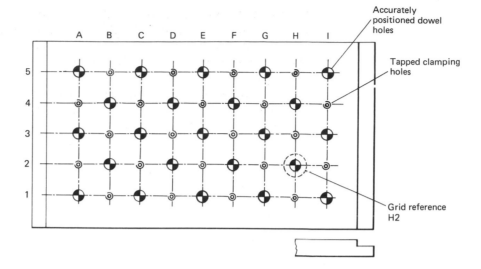

Accurately positioned dowel holes

Tapped clamping holes

Grid reference H2

Fig. 6/21b Grid plate using machined slots

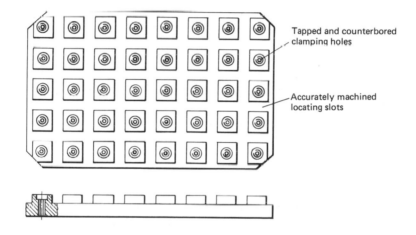

Tapped and counterbored clamping holes

Accurately machined locating slots

within the CNC control unit. In a roundabout way this creates the datum position for the tool tip. The datums positions are not initialised to $X\emptyset$ and $Z\emptyset$ however; they are computed (within the control unit) from the known "limits of travel" dimensions, and the values contained in the axis offset registers.

Regardless of the actual procedure, the part programmer programs the part from a datum point relating to the component. The positioning of the component at the machine, and a similar manoeuvre to those described above, then reconciles the position of the component to the datum positions set at the machine tool. Quite obviously, the correct procedure for each individual machine must be followed. To this end, the operating/setting manual for the individual machine must be thoroughly studied.

Check, and double check, the following points BEFORE initiating an automatic cycle start to machine a component:

1 The component is securely located and clamped.
2 The component and/or workholding device has been positioned in accordance with the instructions supplied by the part programmer.
3 The programmed tool path is not impeded by location or clamping arrangements.
4 The machine datums have been correctly set.
5 The axis offset, and other registers relating to tool position, have been checked to verify their contents. Incorrect values may remain after a component change, for example.
6 The part program has been fully proven (see Chapter 7).
7 All tooling has been secured correctly.
8 All guards are secured correctly in position.
9 The position of the emergency stop button is known.
10 Eye protection is being worn.

Questions 6

1 Discuss the following cutting tool materials and ther applications to CNC machining: HSS, sintered carbides, ceramics, diamond and PCBN.

2 Why are positive rake tooling configurations employed when using sintered carbide cutting tools?

3 Define the term "hardmetal insert" and state any advantages that can be gained by their application in CNC machining.

4 Explain how hardmetal insert tooling should be chosen for any given machining application.

5 Explain the terms "pre-set", "qualified", "semi-qualified" and "indexable insert" in the context of CNC tooling.

6 Explain the concept of a standardised tooling system and why they are useful additions to a CNC machining installation.

7 Describe the application of automatic tool magazines and automatic tool changing on CNC machines.

8 Explain *three* precautions that should be taken when using multiple tools in a programmable turret on a CNC turning centre.

9 What factors influence the choice of speeds and feeds for a given machining operation?

10 An inserted tooth milling cutter is 100 mm diameter and has 12 inserted teeth. If it is to be used at a surface speed of 350 m/min with a feed rate of 0.2 mm/tooth what spindle speed (rev/min) and feed rate (m/min) should be programmed?

11 Why is tool life important in CNC machining installations, and how can it be determined?

12 State the features that an ideal workholding device would possess for CNC applications.

13 What are "bar feeders", "collets" and "face drivers" and when would they be employed in CNC machining?

14 Discuss the principles of location and clamping in the context of the common workholding devices used for CNC work.

15 What is meant by "frictional" and "positive" location?

16 Describe the features and use of a grid plate.

17 Why do some CNC turning centres not have tailstocks?

18 Discuss procedures for setting the workpiece on both CNC Turning Centres and CNC Machining Centres.

19 Discuss the procedures for setting axis datums on both CNC Turning Centres and CNC Machining Centres.

20 State *ten* points that should be checked and double checked before a part program is executed under automatic operation.

Programming considerations 7

7.1 Part programming language

7.1/0 Concept of a stored program

All computer-based systems operate by carrying out sets of instructions in a strict and ordered manner. A collection of such instructions is known as a *program* and the task of providing such instructions is known as *programming*. In the case of a CNC machine tool, such a program is often designed to perform the complete machining of a component or a part. A program of instructions for a CNC machine tool is thus termed a **part program**.

A part program is a set of instructions formulated to describe the actions required of the machine tool. These instructions must be presented to the machine tool control unit (MCU) in the order in which the programmer requires them to be performed. This is because the control program within the MCU is designed to start interpreting the part program from the first line of instructions, in strict sequence, to the last line of instructions. It cannot be stressed enough that that MCU will merely carry out the instructions as presented to it, not what the programmer *thinks* has been presented.

On safety grounds it is important that all instructions, and the sequence of all instructions be checked, and double checked, prior to their execution at the machine. At best, any errors may only result in damage to tooling, the workpiece or the machine tool itself. At worst, the operator could sustain serious injury as a result of a tool/workpiece collision. It is strongly suggested that one, or more, of the methods of proving a part program, presented in section 7.5, be adopted prior to execution at the machine.

7.1/1 Preliminaries to part programming

Part programming should be considered as one of the LAST events to be undertaken in the machining of a component by CNC. Successful machining by CNC methods can be attributed as much to thorough preparatory work, carried out prior to part programming, as it can to the production of the

part program itself. The following points must be considered before the part programming of a component commences:

a) Is the choice of machining by CNC correct for the component under consideration, or can it be accomplished more quickly, simply or cheaply by other means?

b) Has the correct choice of datum(s) and coordinate system been made?

c) Has the machining sequence been correctly planned?

d) Has workholding, setting and clamping been fully considered?

e) Has the choice of tooling, toolchanging sequence and toolchanging position been decided?

f) Have speeds, feeds, coordinate points and other machining data been fully calculated and documented?

g) Has the size of the component been considered in respect of multiple machining (using datum shift, mirror imaging, etc.).

h) Is the programmer fully conversant with the layout, axis movements, operating procedure and facilities of the machine?

It is only when the above points have been fully considered that part programming can commence with absolute confidence. This chapter deals with part programming step by step, and, as such, in comparative isolation. It is presented on the assumption that due consideration has been given to each of the above points.

7.1/2 CNC control systems

When any computer system executes a stored program, that program must conform to two basic conditions. Firstly, it must be presented in a "language" that the system "understands", and secondly the language must be arranged in such a way that it "makes sense". Both conditions must be met. For example, the words—a, you, are, chapter, reading—are all part of our everyday language. However, it is the way that they are arranged that makes them meaningful. For example, they could be arranged as "you are reading a chapter" or "are you reading a chapter", giving two completely different meanings. This is known as the **syntax** of the language. A part programming language is exactly the same. It has a number of **commands** (words) that make up the language, and these must be arranged in a particular **format** (syntax) to be presented to the machine tool control unit.

One difficulty facing students of CNC part programming is that different control units operate with differing commands and differing formats. Thus, any information given on part programming is likely to be accompanied by some conflict and confusion. The intention here is to provide sound principles onto which the peculiarities of any individual control unit may be grafted. No absolute standard yet exists for all CNC programming commands. However, there is a "core" of commands that find almost universal usage among most CNC control units. These commands are drawn from BS3635 Part 1:1972. These commands will be explained and used throughout this chapter. Almost all CNC control units will have additional commands offering more facilities. These commands are left for individual study at the machine in question.

Because of the wide variety of control systems in use, certain assumptions will have to be made throughout this chapter. These will be indicated where appropriate and must be taken into account when considering individual machine tool controllers.

7.1/3 Part program terminology

The MCU controls the machine tool in response to coded commands contained within the part program. These various commands are identified by a capital letter which is referred to as an **address**. This letter address is followed by additional numerical information to make up a command. A command made up of a letter address and its associated numerical information is known as a **word**. A number of words may appear on the same program line. A complete program line is then termed a **block**. Each block is separated by an "end of block" marker.

The agreed order in which the various types of words appear within a block is known as the part program *format*. A uniform format for the blocks of information is usually required for a given control system.

When a part program is provided on punched tape, a single *character* is represented by a single row of holes across the tape width. A *word* comprises several rows of holes depending on the number of characters it contains. A complete *block* of information contains all the rows of holes between two "end of block" characters. This is illustrated in Fig. 7/1.

Fig. 7/1 Part program terminology represented in punched tape

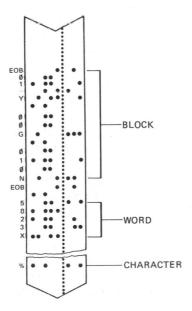

153

7.1/4   Part program codes

N *Block Sequence Number address.* Followed by up to 3 digits, e.g. NØ1Ø, NØ2Ø. In practice most controllers do not follow BS3635 in this respect. With increasing memory capacity in modern CNC control units, a limitation of 999 blocks offered by three digits would prove restrictive for many part programs. Between 4 and 6 digits following the address letter is most common. The program will be executed from the lowest block number to the highest unless instructed otherwise by other program commands. In the traditions of computing it is convention to start from block number 1 and proceed in steps of 5 or 1Ø. Thus, accidentally omitted blocks may be inserted easily without having to re-sequence the entire program.

X, Y, Z *These addresses signify axis motion in accordance with the designated axis motions of the machine tool.* These addresses could be supplemented by **W**, **A**, **B** and **C**, etc. if the machine has extra axes of motion. (See Chapter 2, section 2.2/2 for further details.)

The dimension address will be followed by 7 or 8 characters. Normally 4-digits will be assumed before the decimal point and 2-digits after the decimal point. Both the plus and minus signs are valid characters, normally the plus character being optional. If the position of the decimal point is not assumed by the MCU, then the decimal point character (the period) may also be valid.

Some control systems require the full 7 or 8 digits to be present within the word; others allow leading or trailing zeros to be omitted.

I, J, K *These addresses are used when employing circular interpolation to specify the centre of the programmed arc.* The comments for X, Y and Z coordinate addresses equally apply.

T *Tool Function code identifying the tool to be used, or loaded if at a tool change.* Normally followed by 1 or 2 digits. The tool code range is from TØØ to T99. The programmer must first have assigned each tool a tool code. Accompanying each tool will be a corresponding **tool length offset** which is accessed, during machining, by the tool code. A tool length offset code may have to be specified when using this address, normally by the addition of two extra identifying digits.

S *Spindle Speed letter address.* The digits following the address will be individual to each particular machine. They may refer to a particular speed range, an actual spindle speed, or a code to represent an actual spindle speed. Usually, a 3, 4 or 5 digit code is employed.

F *Feed Rate letter address.* The digits following the address will be individual to the particular machine. The comments for the spindle speed letter address equally apply.

M *Miscellaneous Function letter address.* M-functions are a family of instructions that cause the starting, stopping or setting of a variety of machine functions. The address letter is followed by 2-digits. Some M-functions have been standardised by popular usage and others have special significance for individual machines. The common standardised functions are listed below.

MØØ—Program Stop	(AC)	MØ6—Tool Change		
MØ1—Optional Stop	(AC)	MØ7—Mist Coolant On	(W)*	
MØ2—End of Program	(AC)	MØ8—Flood Coolant On	(W)*	
MØ3—Spindle On—C/W	(W)*	MØ9—Coolant Off	(AC)*	
MØ4—Spindle On—CC/W	(W)*	M1Ø—Clamp On		*
MØ5—Spindle Off	(AC)*	M3Ø—Rewind Tape	(AC)	

The bracketed letters denote the timing of the particular function within the block in which they appear. (AC) indicates that the M-function will be executed After Completion of any commanded axis motion, and (W) indicates that the function will be executed With any commanded motion. An asterisk denotes that the function is retained until it is cancelled or superseded. Such functions are known as **modal functions**.

G   *Preparatory Function letter address*. G-functions are a family of instructions that change the mode of operation of the control. For example, changing from metric to inch units, or from absolute to incremental coordinates. Many G-functions have been standardised and others are individual to particular machines. Common, standardised G-functions are listed below.

GØØ—Rapid Movement	G7Ø—Inch Units
GØ1—Linear Interpolation	G71—Metric Units
GØ2—Circular Interpolation—C/W	G8Ø—Cancel Fixed Cycles
GØ3—Circular Interpolation—CC/W	G81—Series for Fixed Cycles
*GØ4—Dwell	G9Ø—Absolute Coordinates
G3Ø—Series for Mirror Image	G91—Incremental Coordinates
G33—Screwcutting	*G92—Preset Absolute Register
G4Ø—Series for Cutter Compensation	G96—Constant Surface Speed

The above G-functions are modal except those marked with an asterisk. They will be explained in greater detail later.

%   *Program Start Character*. Indicates the start of a part program. It is also used as a "re-wind" stop character when punched tape is employed as an input medium. Note that there is no corresponding single character to indicate the end of the program.

/   *Optional Block Skip Character*. All blocks of information preceded by this character can be caused to be ignored, by the setting of a switch on the console. This is particularly useful for components which are identical other than for certain optional features, such as the presence of holes, or holes in different positions. A "composite" part program may be written and the blocks containing information about the optional features marked with the block skip character. These blocks may then be switched in or out depending on which version of the component is being machined. This character may, in some cases, appear *within* a program block. In such cases it will be interpreted as a data separator.

:   *Alignment Function*. Used to indicate a point in the program at which all operations may be commenced or recommenced. Thus, it may be employed as an optional (or reference) re-wind stop.

()   *Control In and Control Out characters*. Indicates that any part of the program contained within these characters is not to be interpreted by the machine control unit.

∗, #, $, =   These characters can have special significance in the *definition and execution of repetitive programming techniques*. They are likely to be assigned different functions on different control systems. Examples of their likely use are given in section 7.3 in the context of repetitive programming.

7.1/5   Part program formats

The above codes form the "words" of the part program language; we shall now consider the order in which these words are allowed to appear. This is known as the part program *format* and is equivalent to the "syntax" of the language.

The order in which the words appear within a block may be either "fixed" or "variable". The format is then referred to as a **fixed block format** or **variable block format**. There are one or two variations within these format types. The three most common formats will be described.

FIXED SEQUENTIAL FORMAT (Fixed Block)
The instructions in a block are always given in the same sequence. This sequence will be specified by the designers of the MCU. ALL instructions must be given in EVERY block, including those instructions that remain unchanged from the preceding block. Thus, even though an X-dimension may remain constant from one block to the next, it will still have to be specified, in full, in every block. Note that since the words are expected, and always provided, in the same set sequence, the identifying address letters need *not* be provided.

TAB SEQUENTIAL FORMAT (Fixed Block)
The instructions in a block are always provided in the same sequence (similar to the fixed sequential format), and each word is preceded by the TAB character. If instructions remain unchanged in succeeding blocks, the instructions need not be repeated but the TAB character must be punched to ensure that the same number of TAB characters appear in every block. The word address letters need not be employed to identify the words appearing within the block.

WORD ADDRESS FORMAT (Variable Block)
Each word is preceded, and identified, by its letter address. This system enables instructions which remain unchanged from the preceding block, to be omitted from succeeding blocks. This system speeds programming, and tape lengths are considerably reduced. This is the format adopted by most CNC machine control units.

If a control system uses the word address format, the manufacturer may describe the form that the words must take using a **detailed format classification**. An example of such a classification is provided here for completeness:

N3.G2.X42.Y42.Z32.F3.S4.T2.M2

This is interpreted as follows:

N3   Sequence Address N is followed by 3-digits.
G2   Preparatory Functions G are followed by 2-digits.
X42   X-dimensions are followed by 6-digits: 4 preceding the decimal point and 2 succeeding it.
Y42   As X42 but for Y-dimensions.
Z32   Z-dimensions are followed by 5-digits: 3 preceding the decimal point and 2 succeeding it.
F3   Feed Address F is followed by 3-digits.
S4   Speed Address S is followed by 4-digits.
T2   Tool Address T is followed by 2-digits.
M2   Miscellaneous Functions M are followed by 2-digits.

Part programming is normally carried out by using pre-printed **part programming sheets**. These sheets present a neat and ordered way of setting down a part program and, more importantly, provide permanent documentation of the job being machined. Their use should be encouraged.

A typical part programming sheet is shown in Fig. 7/2.

Note that, on the blank part program sheet, there are two lines provided per entry. The handwritten program is put on the upper line. If a *teletype* is then used for tape production, the part program sheet may be sheet-fed into the teletype in the same way as it would be fed into a typewriter. The program may then be copy typed onto the second line, immediately underneath the handwritten program, whilst simultaneously producing the punched tape. This arrangement makes it easy to copy type the part program and also provides a means of checking that the punched tape is identical to the handwritten part program.

A teletype is an electro-mechanical device which performs a similar function to a typewriter. In addition to providing a printed record of whatever is typed, it can simultaneously punch a paper tape. The process can be carried out in reverse in that a punched tape may be read by the teletype to produce a hardcopy (printed listing) of the contents of the punched tape. The teletype still remains the most common method of tape production in CNC installations. Modern teletypes are capable of communicating (both ways) with the CNC control unit direct.

PART PROGRAMMING SHEET

COMPONENT:
CO-ORD DRAWING:
M/C:
DATE:
CUSTOMER:

DATUM INFORMATION:

POSN.	N	X	Y	Z	I	J	K	G	M	T	F	S	OPERATION

SPECIAL INSTRUCTIONS:

PREPARED: SHEET   OF   DRG. No: ISSUE:

Fig. 7/2 A typical part programming sheet

7.2 Part programming procedure 1

7.2/0 Control system assumptions

As a basis for considering the practice of part programming, a control system having the detailed format classification of the preceding section will be assumed. All axes will be assumed to be under single or simultaneous contouring control. Machine datums will be specified assuming a zero offset facility. Programming examples will use Word Address Format allowing only *one* Preparatory (G) function and *one* Miscellaneous (M) function per block of information. For ease of explanation, all dimensional information will be given in millimetres and will *include* the decimal point. Speed codes will be direct rev/min values and Feed codes given in mm/min. In the interests of clarity, each word in the part program segments that follow will be separated by a space. These spaces will not be required when producing part programs on input media such as punched tape.

When milling, all tool movements should be programmed by assuming that it is the tool or cutter that moves, even though in practice it may be the worktable that moves. The reasons for this were explained in Chapter 2, section 2.2/3.

7.2/1 Starting a part program

The **starting point** in a part program is to inform the control system of the various "set-up" conditions for the machining task that follows. Most part programs will start in a familiar, almost identical, fashion. The first few blocks of information should specify, at least:

- The program start character.
- Coordinate values—either absolute or incremental.
- Dimensional units—either metric or inch.
- Tool number (if applicable).
- Spindle speed.
- Start spindle rotation.

Thought must also be given to establishing the datum position of the tool in relation to the workpiece. In many cases the machine can be traversed to the desired position under manual control. The X, Y and Z axes can then be set to zero before the program is run. If this is the case then no provision need be made in the part program. Where the programmer is not the operator, the programmer must ensure that the facility to set a datum position is available to the operator. This can be accomplished by providing a G92 (Preset Absolute Register) code in the first few blocks of the program. Where this facility is available, it will halt execution of the part program and allow the operator to set the values of X, Y and Z as datum values without causing any movement of the machine. The part program may be continued by pressing the start button on the console. Since datum setting only has to be carried out once, this block of information can be marked with the Block Skip character. It can then be ignored on subsequent repeat runs of the program.

Finally, it must be borne in mind that many G-functions are modal, and remain in effect until cancelled or superseded. If part programs are to be run repeatedly (as will be the case in a production environment), it must be ensured that any modal functions in effect at the end of a program run are cancelled prior to it being re-run. For this reason it is often safe practice, at the start of a program, to include function cancel codes, for example G8Ø (cancel fixed cycles) and G4Ø (cancel cutter compensation). It is better practice, however, to cancel any modal functions within the part program after their function has been performed and is no longer required.

The START OF A PART PROGRAM may thus take the following form:

```
%                            .. Program start character
NØØ1  G40                     .. Cancel cutter compensation
NØØ5  G80                     .. Cancel fixed cycle
NØ10  G90                     .. Absolute co-ordinates
NØ15  G71                     .. Metric units
/NØ20 G92 XØ YØ ZØ            .. Dwell to set machine datum
NØ25  F200 S2000 T01 M06      .. Tool change - restate feed and speed
NØ30  M03                     .. Spindle on
```

Each block will be terminated by typing the End of Block character. Note that the comments are merely explanatory notes to guide the reader. They are *not* part of the part program and should not be typed.

NØ25 of the above program segment contains another aspect of good practice, although it may not be immediately apparent. Whenever a tool change is programmed, it is good practice to restate the speed and feed values in effect at that point in the program. The reason is as follows. If a tool breakage occurs during machining, it is desirable to recommence machining (after replacing the broken tool) at the point at which the breakage occurred. Since the machine will have been stopped (and moved to a convenient position) to allow tool replacement, speed, feed and coordinate values will almost certainly be incorrect when restarting. The nearest, previous tool change point will be a convenient place to recommence machining since the coordinate positions, at this point, will almost certainly be known. Supplying feed and speed values within the program will enable the program to be "picked up" without having to start again from the beginning.

> *Note that the above comments may not apply if the machine has been stopped by means of the emergency stop facility. For example, some machines may lose the contents of memory and may automatically move to a home position when the machine is re-started. Knowing the status condition of the machine tool when it is stopped in mid-program—particularly via the emergency stop facility—could be vital.*

If a single tool is used for the complete machining of a component then the tool change manoeuvre may be omitted. The spindle should be turned on *before* any movement of the machine takes place.

The above comments apply equally well to CNC lathes, although references to the third (Y) axis should be ignored.

7.2/2 Programming positional moves

A **positional move** causes the tool or workpiece to move to a commanded position without any cutting taking place. Such moves are characterised by the fact that they are normally done under rapid traverse. They can be programmed under feed control but, since such moves are non-productive (non-cutting), they should be achieved in as short a time as possible. Rapid positioning mode is activated by issuing G00 within the part program.

Machine tool tables commanded to move under rapid traverse can move at speeds in excess of 250 mm/sec. Leaning on machine tool tables, or standing in close proximity to the line of axis movement, must be avoided at all times. Sudden and often unexpected movement of machine tool elements can inflict painful, and potentially serious injury on the unwary.

In the case of a 3-axis milling machine, movement is normally required in the X–Y plane of the machine worktable, and between the current spindle-workpiece position and a coordinate point specified. It is imperative that the machine spindle be retracted to clear any likely obstructions (clamps, workpiece features, etc.) *before* movement commences. A positional move consists of the following:

- Set rapid traverse mode.
- Retract spindle.
- Move in X and Y axes first.
- Spindle down to gauge height.
- Set feed (linear interpolation) mode.

A POSITIONAL MOVE may take the following form:

```
N035 G00 Z0            .. Set rapid traverse, retract spindle
N040 X150.75 Y250.25   .. Move in X-Y
N045 Z-50.50           .. Spindle down to gauge height, rapid
N050 G01               .. Set linear interpolation (feed) mode
```

Although the three axes of a milling machine may be under contouring control, G00 normally means point-to-point positioning. Thus, movement may not be by a straight line between the current position and the point specified. Reference back to Chapter 3, section 3.3/0, will confirm that the path taken is likely to be somewhat irregular.

The term **gauge height** or **gauge plane** is given to mean a reference plane, above the workpiece surface, at which rapid traverse of the approaching spindle (in the Z-axis) must cease and change to feed motion. The nature of the workpiece (a flat plate, for example) may permit retraction of the spindle to this height (rather than Z0) prior to executing positioning moves.

> *It is quite possible to program a positional move, under rapid traverse, in all 3 axes at once. The resulting movement, in 3 dimensions, may not be readily visualised by inexperienced part programmers, so adhere to programming movements in X–Y BEFORE initiating any –Z movements.*

When programming in incremental mode, it should be remembered that measurements for subsequent moves are taken from the last point visited, rather than any previously established datum point.

Although the CNC lathe is free from considerations of a third axis, the zero datum in the Z-axis is often the chuck face. Whilst this may seem an odd choice of datum, it does mean that all absolute programmed moves in X and Z axes will be positive in sign. It would obviously be disastrous to program a move to X∅,Z∅, especially under rapid traverse. Programmers of CNC lathes must be especially vigilant when programming positional moves under rapid traverse since it is extremely easy to command the tool to enter the workpiece envelope, resulting in a serious collision. It is often safer to consider positioning moves in each axis separately, until a working familiarity is gained.

7.2/3   Machining using linear interpolation

Machining using **linear interpolation** simply means machining in straight lines. These lines may be horizontal, vertical, or at an angle, in any direction. All machining is done under control of feed. Linear interpolation mode is entered by issuing a G∅1 code within the part program.

When milling, machining may be commenced in one of two ways. Firstly, with the cutter at the chosen depth of cut, the tool can be programmed to negotiate the workpiece during the interpolated move. Secondly, the cutter may be plunged (or drilled) into the workpiece, to a desired depth of cut, and then the interpolated move initiated. The choice is dictated by the workpiece form and the programmer.

> *Conventional end mills are not suitable for entering the workpiece via a "drilling" operation. A broken tool is the inevitable result of attempting to do so. Where such an approach is required, then slot drills must be employed.*

Linear interpolation operates, in a truly linear way, between the current tool/workpiece position and a commanded coordinate position. Such a move should specify:

- Set linear interpolation mode.
- Set feed rate unless already set.
- Turn coolant on (if required).

- Position spindle (Z-axis move).
- Feed to desired coordinate position (in X and Y).

A corresponding program segment, when milling, to make a 50 mm horizontal cut, followed by a 50 mm vertical cut, 5 mm into the surface of the workpiece (following on from the previous segment) may take the following form:

```
N050 G01 Z-55.50 M08    ..Set interpolation, feed down, coolant on
N055 X200.75            ..Horizontal cut
N060 Y300.25            ..Vertical cut
N065 G00 Z-50.50 M09    ..Retract spindle to gauge height - rapid,
                          coolant off
```

Note that the cut in X and the cut in Y are commanded in separate blocks. Had they appeared in the same block, a diagonal cut would have resulted. If the same operation is carried out using incremental coordinates, then the programmed X and Y values are simply the displacements required, i.e. in this case X5$\emptyset$.$\emptyset\emptyset$ and Y5$\emptyset$.$\emptyset\emptyset$.

When turning, each cut under linear interpolation is often followed by a rapid move, back to the start point, ready to take a subsequent cut. Before returning to the start point, it is practice to retract the tool slightly, from the surface of the workpiece, to avoid scouring the work surface.

> *The datum in X (X$\emptyset$) is always the centre-line of the lathe spindle. Care has to be taken when turning since control systems can differ in the way in which they interpret X-coordinate commands. An X move stated, within a program, as X3$\emptyset$.$\emptyset\emptyset$ could thus mean a finished diameter of 60 mm or, to some control systems "working on diameter", a finished diameter of 30 mm.*

To program a cut along the axis of a turned component, the following approach may be followed:

- Rapid to required depth of cut, 1 mm away from component face.
- Set linear interpolation.
- Traverse length of cut (in Z-axis).
- Retract tool (in X-axis).
- Set rapid traverse.
- Return to start point, rapid.

Program examples in both absolute and incremental coordinates are illustrated in Fig. 7/3.

Always program the cutter path using the centre of the cutter as the cutter path. This is an important principle and will be explained later.

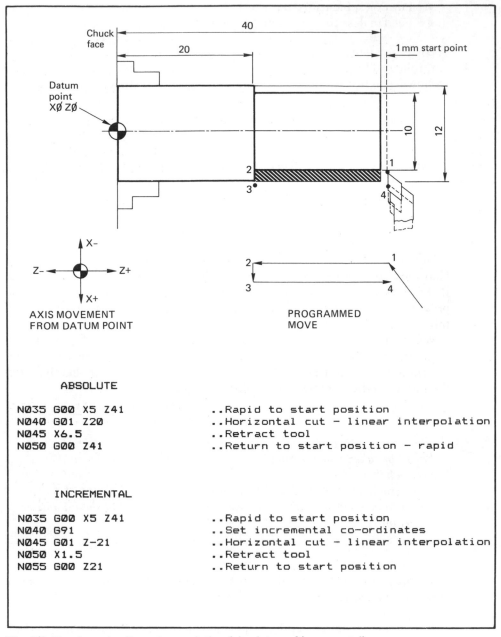

Fig. 7/3 Turning using linear interpolation (absolute and incremental)

7.2/4 Machining using circular interpolation

Circular interpolation means the programming of circular arcs. A single circular interpolation command block is capable of producing a circular arc spanning up to 90°. It is possible to program consecutive blocks to produce half, three-quarter or full circles. It is thus possible to program an arc spanning any number of degrees. Circular interpolation is limited to contouring in a single plane (i.e. in two dimensions only). However, when milling, this plane may be selectable. Figure 3/8a illustrated the effect of employing circular interpolation in the X–Y, X–Z and Y–Z planes of movement. Note that the X–Y plane is the normal plane of the worktable and it is this plane that may be assumed as a default value.

When machining an arc, *four* pieces of information need to be specified:

1) The coordinate position of the START of the arc.
2) The coordinate position of the END of the arc.
3) The RADIUS of the arc to be cut.
4) The DIRECTION (C/W or CC/W) of the cut.

In general, these are specified within the part program, in the following way.

1) The START position is assumed to be the current tool, or cutter, coordinate position.

2) The END position, of the arc, is specified by X, Y and/or Z coordinates measured from the START position.

3) The RADIUS is dealt with by specifying the coordinate position of the centre of the required arc. Since addresses X and Y have already been used to specify the end point, the letter addresses I, J and K are used for this purpose. I is used to specify the centre of the arc in the X-direction; J is used to specify the centre of the arc in the Y-direction; and K is used to specify the centre of the arc in the Z-direction. Control systems differ slightly in the way in which they interpret the coordinates associated with I, J and K. There are two common methods:

Method 1 I, J and K values are dimensions to the centre of the required arc measured from the START POSITION OF THE ARC, in the X, Y and Z directions respectively.

Method 2 I, J and K values are dimensions to the centre of the required arc measured from the PROGRAM DATUM, in the X, Y and Z directions respectively.

4) The DIRECTION of cut is specified by a unique G-code. From BS3635 Part 1:1972, GØ2 is used to specify clockwise circular interpolation and GØ3 is used to specify counterclockwise circular interpolation.

When turning, clockwise and counterclockwise directions are determined by looking onto the top surface of the cutting tool. For inverted tooling this involves looking at the tool from below.

When milling, clockwise and counterclockwise directions depend on the plane in which circulation interpolation is taking place. These planes are selected using G17 (X–Y plane), G18 (Z–X plane), and G19 (Y–Z plane) respectively, before commencing circular interpolation. The directions are shown in Fig. 7/4.

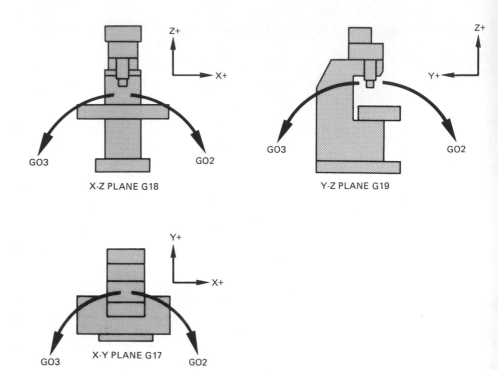

Fig. 7/4 Selectable planes for circular interpolation when milling

X-Z PLANE G18

Y-Z PLANE G19

X-Y PLANE G17

Program segments for illustrating the programming of circular interpolation moves when milling are shown in Fig. 7/5. Assume the spindle to be located at the datum position and retracted to Z∅, 52 mm above the surface of the workpiece. Depth of cut is to be 3 mm into the workpiece.

Similar examples are shown in Fig. 7/6, illustrating the programming of circular interpolation moves when turning. Different control systems will deal differently with the parameters associated with circular interpolation.

The previous examples have dealt with circular interpolation by programming complete 90° quadrants. These moves are characterised by the fact that the start and end points coincide with the centre of the required arc. There are many occasions when the required arc is not a complete quadrant and, as such, the start and end points do not coincide with the arc centre. In such cases the start and end points of the arc have to be calculated. This normally requires the application of trigonometry, geometry, Pythagoras' Theorem, the sine and cosine rules, etc., and a working knowledge of them is thus a desirable attribute of a part programmer (see Section 5.4).

Worked examples of calculations involving partial arc calculations are shown in Fig. 7/7a and b. Note that the coordinates of the start and end points, as calculated, represent positions on the component. These coordinates may need modification if programming to the centre of the cutter (see Section 7.4). It is possible, with some control systems, that there may be constraints concerning the maximum or minimum radius that must be observed.

Many modern control systems are likely to have preparatory functions that will allow for "full circle" circular interpolation programmed within a single program block.

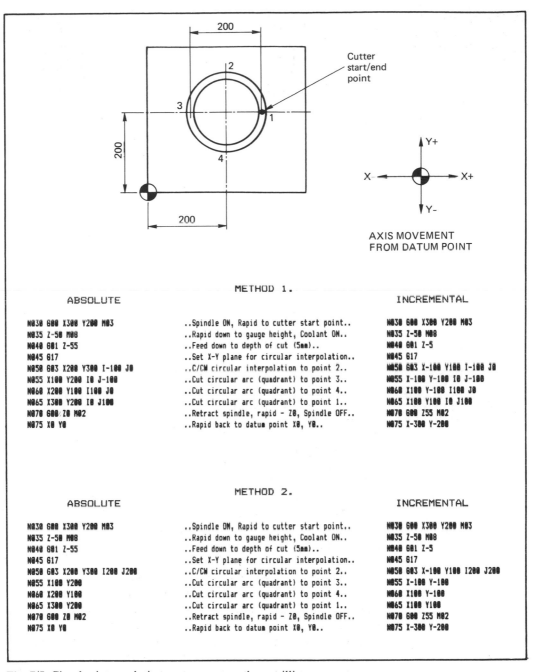

METHOD 1.

ABSOLUTE

N030 G00 X300 Y200 M03	..Spindle ON, Rapid to cutter start point..	
N035 Z-50 M08	..Rapid down to gauge height, Coolant ON..	
N040 G01 Z-55	..Feed down to depth of cut (5mm)..	
N045 G17	..Set X-Y plane for circular interpolation..	
N050 G03 X200 Y300 I-100 J0	..C/CW circular interpolation to point 2..	
N055 X100 Y200 I0 J-100	..Cut circular arc (quadrant) to point 3..	
N060 X200 Y100 I100 J0	..Cut circular arc (quadrant) to point 4..	
N065 X300 Y200 I0 J100	..Cut circular arc (quadrant) to point 1..	
N070 G00 Z0 M02	..Retract spindle, rapid - Z0, Spindle OFF..	
N075 X0 Y0	..Rapid back to datum point X0, Y0..	

INCREMENTAL

N030 G00 X300 Y200 M03
N035 Z-50 M08
N040 G01 Z-5
N045 G17
N050 G03 X-100 Y100 I-100 J0
N055 X-100 Y-100 I0 J-100
N060 X100 Y-100 I100 J0
N065 X100 Y100 I0 J100
N070 G00 Z55 M02
N075 X-300 Y-200

METHOD 2.

ABSOLUTE

N030 G00 X300 Y200 M03	..Spindle ON, Rapid to cutter start point..	
N035 Z-50 M08	..Rapid down to gauge height, Coolant ON..	
N040 G01 Z-55	..Feed down to depth of cut (5mm)..	
N045 G17	..Set X-Y plane for circular interpolation..	
N050 G03 X200 Y300 I200 J200	..C/CW circular interpolation to point 2..	
N055 X100 Y200	..Cut circular arc (quadrant) to point 3..	
N060 X200 Y100	..Cut circular arc (quadrant) to point 4..	
N065 X300 Y200	..Cut circular arc (quadrant) to point 1..	
N070 G00 Z0 M02	..Retract spindle, rapid - Z0, Spindle OFF..	
N075 X0 Y0	..Rapid back to datum point X0, Y0..	

INCREMENTAL

N030 G00 X300 Y200 M03
N035 Z-50 M08
N040 G01 Z-5
N045 G17
N050 G03 X-100 Y100 I200 J200
N055 X-100 Y-100
N060 X100 Y-100
N065 X100 Y100
N070 G00 Z55 M02
N075 X-300 Y-200

Fig. 7/5 Circular interpolation program examples—milling

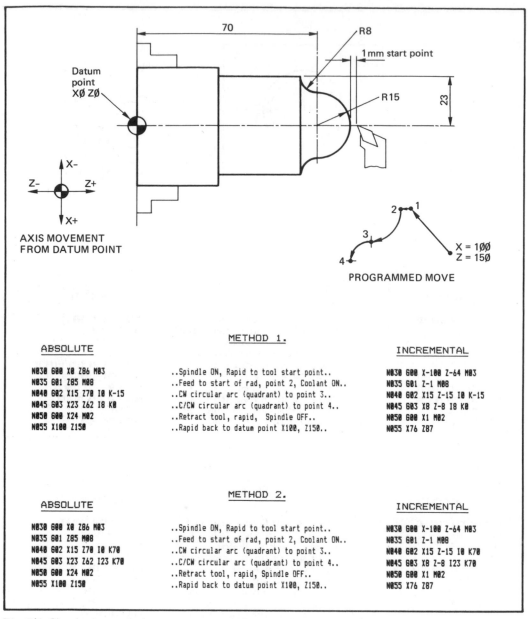

Fig. 7/6 Circular interpolation program examples—turning

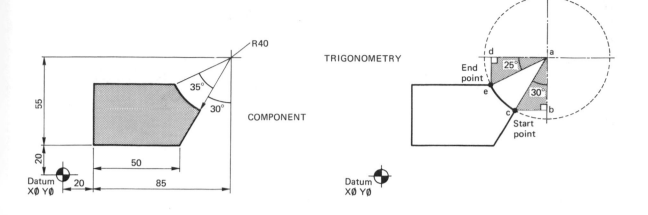

① **To find coordinates of start point C** (Xc, Yc)

Use triangle abc	To find X dimension	To find Y dimension
(triangle diagram: a, 40, 30°, Y1, c, X1, b) $SIN = \dfrac{OPP}{HYP}$ $COS = \dfrac{ADJ}{HYP}$	$\sin 30° = \dfrac{X1}{40}$ $\therefore X1 = 40 \times \sin 30°$ $= 40 \times 0.5$ $\underline{X1 = 20}$ $Xc = 20 + 85 - X1$ $= 105 - 20$ $\therefore \underline{Xc = 85}$	$\cos 30° = \dfrac{Y1}{40}$ $\therefore Y1 = 40 \times \cos 30°$ $= 40 \times 0.8000$ $\underline{Y1 = 34.64}$ $Yc = 20 + 55 - Y1$ $= 75 - 34.64$ $\therefore \underline{Yc = 40.36}$

② **To find coordinates of end point e** (Xe, Ye)

Use triangle ade	To find X dimension	To find Y dimension
(triangle diagram: d, X2, a, 25°, Y2, 40, e) $SIN = \dfrac{OPP}{HYP}$ $COS = \dfrac{ADJ}{HYP}$	$\cos 25° = \dfrac{X2}{40}$ $\therefore X2 = 40 \times \cos 25°$ $= 40 \times 0.9063$ $\underline{X2 = 36.25}$ $Xe = 20 + 85 - X2$ $= 105 - 36.25$ $\therefore \underline{Xe = 68.75}$	$\sin 25° = \dfrac{Y2}{40}$ $\therefore Y2 = 40 \times \sin 25°$ $= 40 \times 0.4226$ $\underline{Y2 = 16.90}$ $Ye = 20 + 55 - Y2$ $= 75 - 16.90$ $\therefore \underline{Ye = 58.10}$

③ **To find coordinates of arc centre I, J**

METHOD 1		METHOD 2	
Arc centre positioned relative to start point of arc		Arc centre positioned relative to datum point X0, Y0	
I dimension	J dimension	I dimension	J dimension
I = X1 see above calculation $\therefore \underline{I = 20}$	J = Y1 see above calculation $\therefore \underline{J = 34.64}$	I = 20 + 85 from given dimensions $\therefore \underline{I = 105}$	J = 20 + 55 from given dimensions $\therefore \underline{J = 75}$

Fig. 7/7a Partial arc calculations—milling

169

COMPONENT

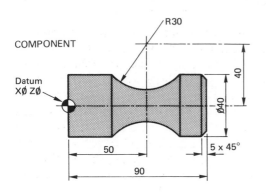

TRIGONOMETRY

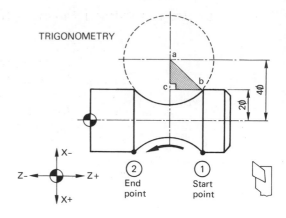

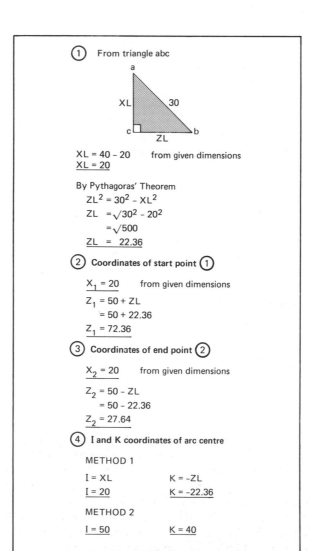

① From triangle abc

XL = 40 – 20   from given dimensions
__XL = 20__

By Pythagoras' Theorem
$ZL^2 = 30^2 - XL^2$
$ZL = \sqrt{30^2 - 20^2}$
$= \sqrt{500}$
__ZL = 22.36__

② Coordinates of start point ①

$\underline{X_1 = 20}$ from given dimensions
$Z_1 = 50 + ZL$
$= 50 + 22.36$
$\underline{Z_1 = 72.36}$

③ Coordinates of end point ②

$\underline{X_2 = 20}$ from given dimensions
$Z_2 = 50 - ZL$
$= 50 - 22.36$
$\underline{Z_2 = 27.64}$

④ I and K coordinates of arc centre

METHOD 1

$I = XL$ $K = -ZL$
$\underline{I = 20}$ $\underline{K = -22.36}$

METHOD 2

$\underline{I = 50}$ $\underline{K = 40}$

Fig. 7/7b Partial arc calculations—turning

7.2/5 Part programming using fixed or canned cycles

A **canned cycle** or **fixed cycle** is a fixed sequence of operations, inbuilt to the control system, that can be brought into action by a single command. Such cycles are defined in order to considerably reduce programming time and effort, on repetitive and commonly used machine operations. The adoption of such canned cycles can also considerably reduce program length.

Canned cycles form part of the family of preparatory G-functions. G81 to G89 are reserved for the various cycles and G80 used as the Fixed Cycle Cancel function. G80, in addition to cancelling any fixed cycle, will usually position the tool/workpiece and return the spindle to gauge height. Canned cycles are modal functions and remain operative until superseded by a subsequent canned cycle, or cancelled by G80 (and/or G00 depending on the control system).

To illustrate the use of these cycles we shall first consider the action of drilling a hole under program control. The sequence for such an operation is:

- Move spindle to gauge height, turn spindle on.
- Move, in rapid, to the centre coordinate of the hole.
- Feed to required depth.
- Rapid out to gauge height ready for next move.

A part program segment for achieving this might be as follows. Assume a hole to be drilled at coordinate position X100Y100, gauge height to be at Z-50 and the required depth 5 mm. Thus, in absolute:

```
N050 G00 Z-50 M03      ..spindle to gauge height, rapid, spindle ON
N055 X100 Y100         ..rapid to hole co-ordinate position
N060 G01 Z-55 M08      ..feed to depth, coolant ON
N065 G00 Z-50          ..rapid out to gauge height
```

When programmed using a fixed drilling cycle G81, the program segment becomes:

```
N050 G00 Z-50 M03          ..spindle to gauge ht., rapid, spindle ON
N055 G81 X100 Y100 Z-55 M08 ..move to co-ordinate position, drill to
                              depth, retract to gauge ht., rapid
```

This program segment has been reduced by half by utilising the G81 fixed drilling cycle. This could significantly reduce the length of the part program if many similar holes have to be drilled. Fixed cycles automatically perform a number of discrete operations as designated by the appropriate G-code. A convention has been adopted for illustrating the actions performed by the various fixed cycles. This schematic convention is shown in Fig. 7/8. Note that the fixed cycle G-code must be accompanied by additional information in the same block. The common fixed cycles are represented in Fig. 7/9.

Fig. 7/8 Schematic convention illustrating the action of fixed cycles

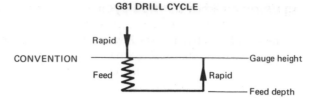

G81 DRILL CYCLE

CONVENTION

Rapid

Feed

Gauge height

Rapid

Feed depth

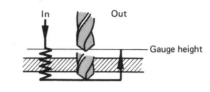

SITUATION

In

Out

Gauge height

Fig. 7/9 Schematic representation of the common fixed cycles

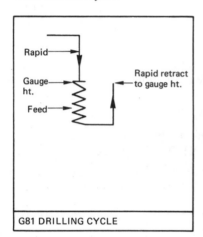

Rapid

Gauge ht.

Feed

Rapid retract to gauge ht.

G81 DRILLING CYCLE

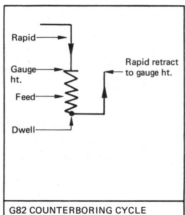

Rapid

Gauge ht.

Feed

Dwell

Rapid retract to gauge ht.

G82 COUNTERBORING CYCLE

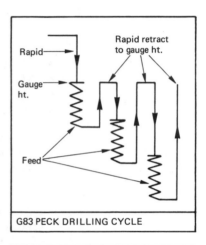

Rapid

Gauge ht.

Feed

Rapid retract to gauge ht.

G83 PECK DRILLING CYCLE

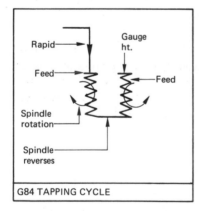

Rapid

Feed

Spindle rotation

Spindle reverses

Gauge ht.

Feed

G84 TAPPING CYCLE

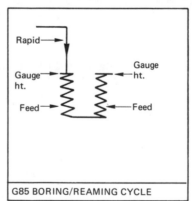

Rapid

Gauge ht.

Feed

Gauge ht.

Feed

G85 BORING/REAMING CYCLE

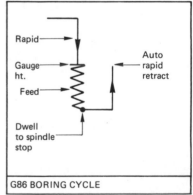

Rapid

Gauge ht.

Feed

Dwell to spindle stop

Auto rapid retract

G86 BORING CYCLE

Fixed cycles for turning operations, whilst widely available, are somewhat difficult to describe. Common fixed cycles include straight turning, taper turning, face turning, and taper face turning cycles; area clearance (stock removal), grooving and peck drilling cycles; thread turning, taper thread turning, and multi-start thread turning cycles. Most cycles must be accompanied by additional information in the command block. Thread turning, for example, may require the lead, depth of thread, and number of passes to be specified.

A stock removal, or area clearance cycle, is simply a roughing cycle whereby a number of passes are made to clear large amounts of material. The tool will traverse a similar tool path, automatically increasing its depth of cut at each pass. The information that must accompany the fixed cycle specification can be specified in so many different ways that it would not be helpful to offer specific program examples.

All fixed cycle G-codes must be accompanied by additional information within the commanded part program block.

7.3   Part programming procedure 2

7.3/0   Repetitive programming

The previous section outlined the fundamentals of simple part programming procedure and practice. As the components to be programmed become more complex, the resulting part programs become proportionately longer. Furthermore, if the same or similar features occur within the same component (such as an identical pattern of holes repeated at different positions), it can be a source of much tedium to have to repeat identical sections of the same part program code to produce the repeated features. A more elegant way of achieving such repetition is provided by three constructs available to the part programmer. The utilisation of repetitive programming techniques, by imaginative part programmers, can significantly reduce the length of the resulting part program and dramatically shorten part program development time. Study of the repetitive programming facilities offered by an individual control system will be amply repaid.

The three common facilities are referred to as *loops*, *subroutines* and *macros*. They all operate by altering the flow of the part program according to a set of pre-defined rules. They will be discussed in turn in the following sections. Although these facilities may be available on most CNC machines their implementation is likely to differ slightly on different machines.

7.3/1   Loops

Looping provides the programmer with the ability to jump back to an earlier part of the program and execute the intervening program blocks a specified number of times. It is so called because the program loops back on itself. This is particularly useful when used in conjunction with incremental programming.

In order to repeat a section of the part program a number of times it is necessary to specify *three* pieces of information:

- The START of the loop.
- The END of the loop.
- The NUMBER of REPEATS of the loop.

All three pieces of information are provided in a single block of information like so

$$=N1\emptyset\emptyset/3$$

The = character marks the START of the loop, the N1$\emptyset\emptyset$ denotes the block number where the loop is to END, and /3 specifies the NUMBER of REPEATS (in this case 3).

Suppose that we require to drill 6 holes along a horizontal line (at Y1$\emptyset\emptyset$), each at a spacing of 30 mm from each other, starting from X1$\emptyset\emptyset$. Using the G81 fixed (drilling) cycle, the program segment will be as follows:

```
N030 X70 Y70 M03            ..move, rapid, to X70 Y70, spindle ON
N035 Z-50 M08               ..spindle down to gauge ht., rapid, coolant ON
N040 G91 Y30                ..set incremental co-ordinates, move up Y30
=N050/6 ──────────────┐     ..repeat up to N050 six times
N050 G81 X30 Z-5 F400─┘     ..move along X30, drill Z-5 deep, retract
N055 G80                    ..cancel fixed cycle
N065 G90 Z0 M09             ..retract spindle, absolute, coolant OFF
```

The beauty of the looping structure (indicated by the square bracket) is that any number of holes can be accommodated by simply changing the Number of Repeats value.

When the loop start statement is encountered, the number of repeats is placed into a special memory location called a *register*. After every repeat this register is decremented by 1. If the contents of this register is greater than $\emptyset$, when the program reaches the End of Loop block, then the program will loop back. When it reaches $\emptyset$, the program will "fall through" to the block following the End of Loop block.

The flexibility of the looping structure is further enhanced by the ability to place a loop within a loop. Such structures are then known as **nested loops**. Suppose that we now require 6 rows of the 6 holes we have already programmed, forming a 6×6 matrix of holes. It is possible to place a loop around the outside of the loop structure already present, and repeatedly increment the spindle position (in Y), after 6 horizontal holes have been drilled. The resulting part program segment is shown below.

```
N030 X70 Y70 M03            ..move, rapid, to X70 Y70, spindle ON
N035 Z-50 M08               ..spindle down to gauge ht., rapid, coolant ON
=N060/6 ──────────────────┐ ..repeat up to N060 six times (outer loop)
N040 G91 Y30              │ ..set incremental co-ordinates, move up Y30
=N050/6 ──────────────┐   │ ..repeat up to N050 six times  (inner loop)
N050 G81 X30 Z-5 F400─┘   │ ..move along X30, drill Z-5 deep, retract
N055 G80                  │ ..cancel fixed cycle
N060 G90 X70──────────────┘ ..set absolute, move to X70
N065 G90 Z0 M09             ..retract spindle, absolute, coolant OFF
```

We now have a program segment capable of drilling a matrix of any number of holes, at any pitch, merely by altering one or two values within the existing part program.

There is a limit to the number of repeats allowed, and on how deep (how many levels) loops may be nested. Note that, when utilising nested loops, any inner loops must be fully enclosed (nested) by any outer loops. It is illegal (to the MCU) to have loops that cross each other's boundaries (see Fig. 7/10). There may be other rules to be observed when describing loop structures; the programmer's manual for the individual machine must be consulted.

Fig. 7/10 Legal and illegal nested loop structures

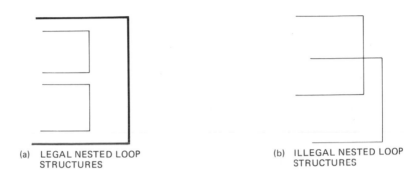

(a) LEGAL NESTED LOOP STRUCTURES

(b) ILLEGAL NESTED LOOP STRUCTURES

It was mentioned earlier that area clearance cycles, when turning, are available as a preparatory G-function. It is easy to relate that this is an application of a simple looping structure. It should further be evident that repetitive programming techniques enable the part programmer to design non-standard fixed cycles.

7.3/2 Subroutines or sub-programs

There may be repetitive features within a component that cannot be accommodated within a loop structure. In such cases the repetitive elements may be described in terms of a **sub-program**, often called a **subroutine**, and placed at the end of the main body of the part program. Whenever the feature is required, within the part program, its associated subroutine is called for execution. When a subroutine is called, the flow of the program is transferred to the start of the subroutine to continue execution. After the subroutine is completed, the flow of the program returns to the main body of the part program at the point immediately following that at which the subroutine was called.

Thus, a subroutine may be called from many different positions within a part program. Note that a subroutine must not be defined within the main body of a part program, but must be defined and placed at its end. To describe, and use, a subroutine, it must be possible to indicate:

- The identification (START) of the subroutine.
- The END of the SUBROUTINE definition.
- The END of the PROGRAM—after which subroutines may follow.
- A means of CALLING a specified subroutine.

For the purposes of illustration, the # (hash) character followed by a number will be used to identify the START of a particular SUBROUTINE. For example, #3 will indicate the start of subroutine number 3. The $ (dollar) character will be used to indicate the END of a SUBROUTINE, and the miscellaneous function M$\emptyset$2 will serve as the End of Program marker. All subroutines must be defined AFTER the end of the End of Program marker. This is to ensure that the flow of the part program does not accidentally enter the subroutine definition and execute it inadvertently. Once defined, a subroutine may be CALLED from the main body of the program by a calling line such as = #3, which would be translated as "call subroutine number 3". A legal call (such as = #3) should be the only means of executing a subroutine.

Consider that it is required to mill a square profile 50 mm × 50 mm, at a depth of 3 mm, at various positions in a flat plate component. A subroutine, which will execute at the current spindle position, might be defined as follows. Gauge height is at Z-5$\emptyset$, which is 2 mm above the surface of the component.

```
#1                      ..start of subroutine number 1.
N100 G01 Z-55 F400      ..feed to required depth, absolute
N105 G91 X50            ..set incremental, traverse 50mm in X+
N110 Y50                ..traverse 50mm in Y+
N115 X-50               ..traverse 50mm in X-
N120 Y-50               ..traverse 50mm in Y-, back to start position
N125 G91                ..set absolute
N130 G00 Z-50           ..retract spindle to gauge ht., rapid
$                       ..end of subroutine definition
```

To produce four such profiles, the above subroutine can be called from the main body of the part program as follows:

```
N030 G90 M03            ..set absolute, spindle ON
N035 G00 X50 Y50        ..move, rapid, to position 1.
N040 Z-50 M08           ..spindle down to gauge ht., coolant ON
=#1                     ..Call subroutine number 1.
N045 X150 Y50           ..move, rapid, to position 2.
=#1                     ..Call subroutine number 1.
N050 X150 Y150          ..move, rapid, to position 3.
=#1                     ..Call subroutine number 1.
N055 X50 Y150           ..move, rapid, to position 4.
=#1                     ..Call subroutine number 1.
N060 Z0 M09             ..retract spindle, rapid, coolant OFF
N065 X0 Y0 M05          ..move to datum point, rapid, spindle OFF
```

When defining subroutines, it is good practice to ensure that the machine is returned to the same condition upon leaving the subroutine as it was when the subroutine was called. For example, it is easy to enter a subroutine under absolute coordinates, and return from the subroutine in incremental mode. There is a limit to the number of subroutines that may be defined within a single part program, and certain other specific rules may have to be observed in their definition. Generally, subroutines may call other subroutines, but subroutines must not be defined within subroutines.

7.3/3 Macros

The term **macro** is short for *macro command* or *macro sub-program*. Strictly speaking, a macro is a single command that generates a series of tool path moves for the execution of a particular machining operation. The macro call, within the part program, must invariably be accompanied by additional information. Macros may be pre-defined as in-built control system facilities, or they may be user-defined. In the latter case, the implementation of a macro is very similar to that of a subroutine. In fact some manufacturers use the two terms synonymously and without distinction.

For our purposes a *macro* can be thought of as *a subroutine with the ability to pass values or parameters*. A **parameter** is a value that acts like a constant value when in use, but may take on different values from run to run. The subroutine described earlier will mill a simple square profile 50 mm × 50 mm. It cannot mill a profile of any other dimensions. If true macro facilities are available it will be possible to pass, with the calling line, parameter values for the dimensions of the sides of the profile to be milled.

For simplicity, it will be assumed that the macro is defined in exactly the same way as the subroutine from the previous section. In addition it is necessary to specify:

- The PARAMETERS that may change, from call to call.
- A means of PASSING true values into these parameters.

Within the macro definition, the parameters likely to change are given no numerical value, but are marked with an asterisk *. True numerical values for these parameters are then supplied when the macro sub-program is called. Different values may thus be supplied with different calling lines.

To illustrate the macro concept, the previous subroutine will be re-constructed as a macro specifying the length and width of the profile to be milled as parameters:

```
#1                      ..start of macro number 1.
N100 G01 Z-55 F400      ..feed to required depth, absolute
N105 G91 X*             ..set incremental, traverse X+ value to be supplied
N110 Y*                 ..traverse Y+ value to be supplied
N115 X*                 ..traverse X- value to be supplied
N120 Y*                 ..traverse Y- value to be supplied, back to start
N125 G91                ..set absolute
N130 G00 Z-50           ..retract spindle to gauge ht., rapid
$                       ..end of macro definition
```

In the calling line, the parameters are passed IN THE ORDER IN WHICH THEY APPEAR in the macro definition. In this example, the calling line must supply values for X(+), Y(+), X(−) and Y(−) in that order. The parameters are identified with the appropriate address letter and the identifying asterisk. A calling line takes the following form:

```
=#1  X*50  Y*50  X*-50  Y*-50
     (X+)  (Y+)  (X-)   (Y-)
```

Note the similarity with a subroutine call but with the addition of actual values for the identified parameters.

If the 4 calling lines in the subroutine example were replaced by 4 lines of the above macro call, an identical component will be machined. The power of the macro call, however, resides in the ability to change the value of the parameters from call to call.

For example, a call to the same macro to mill a rectangle measuring 100 mm × 50 mm takes the following form:

```
=#1  X*100, Y*50, X*-100, Y*-50
     └─┬─┘  └─┬─┘ └──┬──┘  └─┬─┘
      (X+)   (Y+)   (X-)    (Y-)
```

We now have a macro sub-program which machines rectangular or square profiles to dimensions that can be specified either within a program, or edited in at the machine. Furthermore, such flexibility is accomplished within extremely compact part program code. Because parameters may be defined and altered, this technique of programming is also referred to as **parametric programming**.

The above techniques offer considerable scope and flexibility to the part programmer, especially when they are combined together. Although they may, at first sight, appear complicated, their mastery will be amply rewarded by savings in time, greater flexibility in actual machining, and the satisfaction of producing elegant and compact part programs. Taking the above concepts even further, it is possible to define "standard" often-used subroutines and macros as **library** features. Such features may then be added to part programs under development, in a modular fashion. Again it must be stressed that the above implementations are merely illustrative examples of the techniques outlined. Different implementations and further constraints will almost certainly be imposed by individual control systems.

7.3/4 Additional features

Many control systems offer additional features to ease the programming of repetitive features, within a component. Most of these additional features are invoked by specifying preparatory G-codes within the part program. Such codes will be specific to each control system depending on the facilities it offers.

The more common facilities available on CNC milling machines are described below.

REFLECTION   The nature of many engineering products requires that they be supplied in "handed pairs", that is a right-handed component and a left (opposite)-handed component. The two components will invariably be identical in dimension but geometrically opposite. This facility is often known as **mirror imaging** since its effect can be visualised by placing a mirror adjacent to an object and comparing the actual object with the image produced in the mirror. The line along which the mirror is placed is known as the *axis of symmetry* or the *axis of reflection*. On CNC control systems, two axes of reflection are normally accommodated, reflection in the X-axis and reflection in the Y-axis.

A moment's thought will confirm that this is relatively simple to implement. It is necessary only to reverse the signs of all dimensions in the axis of symmetry. All +X dimensions become −X dimensions for reflection in the Y-axis, and all +Y dimensions become −Y dimensions for reflection in the X-axis. The control system automatically arranges this on receipt of the requisite G-code within the part program. Three possibilities are available: reflection in X, reflection in Y, and reflection in both X and Y simultaneously. Note that reflection normally occurs about the absolute programmed datum XØ, YØ.

The effects achieved by reflection are illustrated in Fig. 7/11.

Fig. 7/11 Component reflection or mirror image programming facility

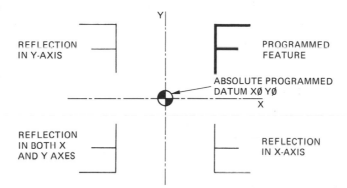

ROTATION   Many components comprise features which consist of a single element that is repeated by **rotating** it about a specified origin. This origin is normally the absolute programmed datum XØ,YØ. Common examples include equi-spaced holes or slots on a common pitch circle diameter, or components comprising radial features such as arms or spokes. If such a facility is specified using a preparatory G-code, it must be accompanied by information specifying the angle of rotation. The programmed feature may be rotated by a specified (absolute) number of degrees, or if used in conjunction with a looping technique, rotated by an incremental amount each time the program is looped.

The effects of rotation are illustrated in Fig. 7/12.

Fig. 7/12 Component rotation programming facility

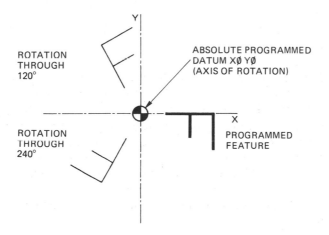

TRANSLATION   **Translation** refers to producing an identical feature (in size, shape and orientation) merely shifted by a specified amount. This shift may be in the X-axis, the Y-axis, or both. This merely requires that the datum start point may be shifted by the required amount. This can be accomplished by utilising the appropriate preparatory G-code within the part program, and specifying the X and Y coordinates for the new start point. Translation is illustrated in Fig. 7/13.

Translation facilities can become very powerful when used in conjunction with loop structures, subroutines and macros, and when employing incremental coordinate moves.

On turning centres it is common to have a pattern repeat cycle. Whilst this is not true translation it allows for the rough and finish machining of preformed parts such as castings or forgings.

Fig. 7/13 Component translation or datum shift programming facility

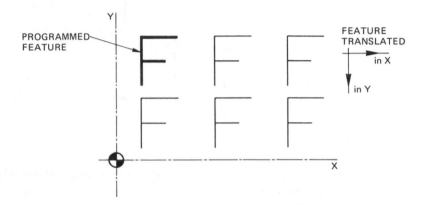

SCALING   This facility allows the cutter path to be transformed by increasing (**scaling up**), or decreasing (**scaling down**) its magnitude relative to its programmed dimensions. The actual programmed dimensions are considered to have a scale value of 1. When the scaling facility is invoked, extra information is supplied specifying the scaling required in the X-axis and similarly in the Y-axis. A family of different-size components may thus be machined from a single part program via a simple edit of the scaling factors. The effect of scaling is illustrated in Fig. 7/14.

Fig. 7/14 Component scaling programming facility

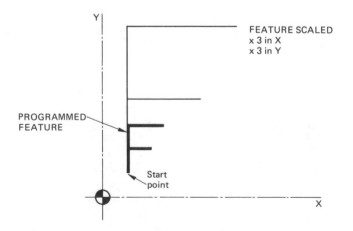

Some CNC machine tools, especially turning centres, may offer the facility of **safety crash zone programming**. This facility allows the operator to set up a work "envelope", in space, which the tool is not allowed to enter. It is a sort of electronic safety zone which prevents the tool from being programmed into potentially dangerous areas. It can be most reassuring to know that the tool can be prevented from negotiating a revolving chuck, for example. Such features should be utilised to the full, even though they may take a little extra time to set up. Some machine tools allow the definition of "safe", "warning" and "prohibited" zones.

Most CNC turning centres have the facility for **constant surface speed** or CSS programming. For a given material (cutting speed), the speed of the spindle depends upon the diameter being cut. (This is explained fully in Chapter 6, Section 6.2/1.) This means that a different speed setting is required for all cuts other than when turning plain diameters. It is impractical to arrange for changes in spindle speed, within the program, whenever the diameter of cut changes. The CNC control system accomplishes this, however, when the CSS option is programmed. This facility is made possible by the use of continuously variable speed motors used for spindle drives. It is usual to specify a maximum spindle speed before calling for CSS, and this will limit the maximum speed to this value. This maximum speed is normally the maximum speed of the machine. It may be desirable to limit the maximum attainable speed, for example when employing faceplates or turning eccentric components. Limiting the maximum speed reduces any vibrations generated by out-of-balance forces. It is characteristic (and often somewhat unnerving) of CNC turning centres to hear the spindle speed increase continuously when taking facing cuts.

With the CSS facility programmed, the nearer the tool to X$\emptyset$, the faster the spindle speed. As the tool retracts from X$\emptyset$, the spindle speed reduces. This can be something of a disadvantage if, for example, a tool change is to be made and the tool has to be moved to a tool change position well away from X$\emptyset$. A rapid deceleration occurs during this move and reduces the efficiency of the part program. The CSS facility should be disabled under such circumstances to restore the priority of rapid traverse.

7.4 Part programming and tooling

7.4/0 Choice of tooling

When planning a part program, decisions must be made regarding the type and size of cutter(s) to be used. Usually, such decisions will be based on the standard tooling carried by the organisation. The part programmer works on **nominal cutter size**, not knowing the actual size and condition of the tools available to the shop floor. Part programs are thus produced on the assumption that nominal size tooling is to be used. In practice, however, nominal size cutters may not be available. Dimensional variations may be present due to factors such as tool wear, tool re-grinds, tool refurbishment after breakage, variations in tool holding devices, etc. Quite obviously if an oversize or undersize cutter is used to machine a component, programmed for a nominal size cutter, then the resulting component will be dimensionally incorrect. Discrepancies may occur both in tool length and tool diameter, and also, in the case of lathe tools, tool width and tool nose radius.

All CNC control systems have facilities to compensate for such discrepancies without affecting the part program. Such facilities operate by specifying the difference between the actual cutter size and the programmed cutter size, before machining takes place.

7.4/1 Tool length offset

A tool *offset* is defined as a correction parallel to a controlled axis. In the case of a CNC *milling* machine such a correction is required to cater for differences between the programmed length of a cutter, and its actual length. Such an offset is known as a **tool length offset** (TLO) and is operative in the Z-axis of movement.

Mention has already been made of the gauge, or reference, plane. This plane is established, by the part programmer, to be just above the surface of the workpiece. The gauge plane has previously been defined as the point (in the Z-axis) at which rapid traverse of the tool reverts to feed. More importantly, the gauge plane normally represents the datum level, about which all movements in the Z-axis are referenced.

When using a single tool for the machining of a complete part, it is common practice to feed the tool down manually to the gauge plane, and set this position to be Z∅. The exact position of the reference plane, above the surface of the workpiece, may be set in one of two ways. The tool may be fed down to touch the surface of a reference or gauge block placed on the workpiece. Alternatively, the tool may be fed down to touch the surface of the workpiece and then retracted by the desired amount, by referencing the Z-axis display on the operator console. In this case (using only one tool) there will be no tool length offset, or more correctly, a tool length offset of ∅.

When a second (or subsequent tool) is employed, or the original tool is replaced, within the same part program, there will almost certainly be a difference in length from that of the original tool. Thus, the effect will be that a longer tool will exhibit deeper depths of cut, and a shorter tool will not cut deep enough. To account for such discrepancies, the differences in length

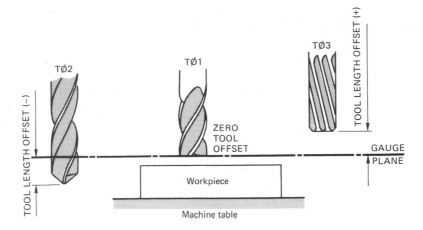

Fig. 7/15 Tool length offsets for milling related cutters

between the first tool and the subsequent tools are noted, and stored as a tool length offset, within the memory of the control unit. Thus, when a different tool is called for use, its corresponding tool length offset causes the programmed Z-axis positions to be modified by this amount. Tool length offsets essentially represent addition or subtraction, by the control unit. Thereafter, the original programmed Z-values will be valid regardless of the tool length being employed.

Tool length offsets, for milling related cutters, are illustrated in Fig. 7/15.

In the case of CNC *lathes*, similar considerations apply, although the relative position of the tools can now vary in both X and Z axes. If the toolholder itself is used as a datum, different tool tips, by virtue of their mounting arrangements, lie at different positions relative to the fixed datums. Thus, when turning, two offsets apply to each tool. The effect of calling a turning tool with associated TLO values is to correct the programmed X and Z axis positions by these amounts. A TLO value is thus likely to have two or three components: an X-offset value, a Z-offset value, and a tool nose radius compensation value. Tool length offsets for turning tools are illustrated in Fig. 7/16.

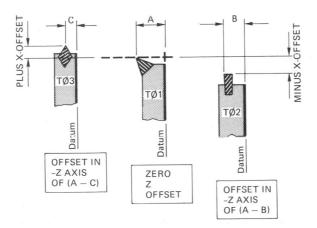

Fig. 7/16 Tool length offsets for turning related cutters

7.4/2 Cutter diameter compensation

It was stated earlier that an important principle of part programming is to program the required cutter path with reference to the centre of the tool rather than at the point on its periphery where actual cutting takes place. This is suggested for three primary reasons:

a) Programming of the cutter path will be considerably simplified.
b) It may be difficult on some cutting tools to ascertain where cutting is actually taking place (in 3-dimensional contouring, for example) and possible errors may result.
c) Differences in the programmed diameter and actual diameter of the cutting tools used can be easily accommodated using cutter compensation facilities.

These points are illustrated in Fig. 7/17*a*, *b* and *c*.
A cutter *compensation* is defined as a correction normal (at right angles) to a controlled axis.
If the principle of programming to the centre of a cutting tool is adopted, then it is clear that the diameter of the cutting tool must be assumed at the time of programming. This will normally be of a standard nominal size. Reference to Fig. 7/17*c* will confirm that, if a smaller-diameter cutter is used, then a correspondingly larger workpiece will result. Conversely, if a larger diameter cutter is used, then the resulting workpiece will be smaller than programmed. Such discrepancies are accounted for by applying **cutter diameter compensation**.
When the actual cutting tools have been chosen, the difference between the actual diameter and the programmed diameter is entered into the control unit via MDI. Each tool will have its own corresponding compensation value. When cutter compensation is initiated (when machining), the control unit generates a new tool path. This new tool path will, at all times, be equidistant from the programmed cutter path. This "new" cutter path will be separated from the programmed cutter path by half the difference between the actual cutter diameter and the programmed cutter diameter. This is the compensation dimension. It will also be necessary to indicate whether any corrections made are to take place to the right, or to the left, of the tool when machining. This will be specified by issuing a preparatory G-code, normally from the G4∅ series. BS3636:Part 1 specifies G41 as compensation applied to shift the program path to the LEFT of the cutter and G42 as compensation applied to shift the program path to the RIGHT of the cutter. G4∅ will cancel cutter compensation. Left and right are identified by assuming a point on the top face of the tool facing in the direction of cut.
Consider Fig. 7/17*c* and the case of using an oversize cutter. If the direction of cut were programmed in a clockwise direction, then cutter diameter compensation would be applied to the left of the cutter. If the direction of cut were programmed in a counterclockwise direction, then cutter diameter compensation would be applied to the right of the tool.
In the case of CNC lathes, all turning tools and cutting tool tips have nose radii (however small) to impart strength and rigidity, and to increase tool life. Standard, replaceable carbide tips have standard radii of 0.5 mm, 0.8 mm,

Fig. 7/17a Calculation of cutter path simplified by programming cutter centre

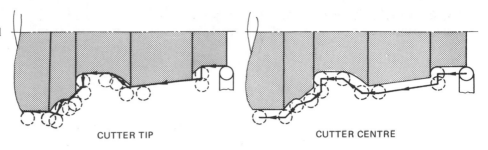

CUTTER TIP

CUTTER CENTRE

Fig. 7/17b Difficult to ascertain where cutting takes place on some cutters

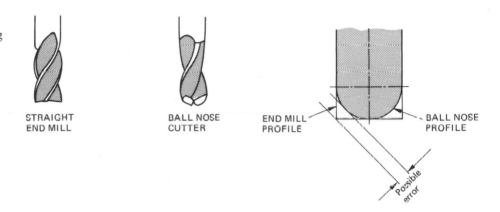

STRAIGHT END MILL

BALL NOSE CUTTER

END MILL PROFILE

BALL NOSE PROFILE

Possible error

Fig. 7/17c Difference in programmed and actual cutter sizes will give incorrect size components

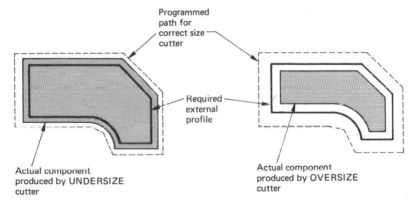

Programmed path for correct size cutter

Required external profile

Actual component produced by UNDERSIZE cutter

Actual component produced by OVERSIZE cutter

1.2 mm or 1.5 mm. Button turning tools can have nose radii as large as 25 mm. This means that, although the tool point exists as a programmable point, it may not exist physically (see Fig. 7/18). This discrepancy, however small, must be accounted for if accurate components are to be produced. This is especially important when employing qualified tooling. There will be no discrepancy in size when turning plain diameters or when straightforward facing cuts are taken. The discrepancy becomes noticeable when machining tapered and/or circular features. It is only strictly necessary to apply cutter compensation when taking finishing cuts. The principles of cutter diameter compensation equally apply although, in this context, it may be referred to as **tool nose radius compensation** or TNR compensation. TNR offset values are provided for small variations in tool point locations.

185

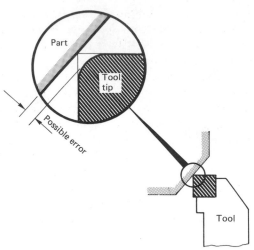

Fig. 7/18 Tool nose radius compensation: the tool point exists as a programmable point but may not exist physically

Where large discrepancies are present (for example, where a number of tools must be used together within the turret), a similar facility is often provided. Such a facility is known as **tool geometry offset**.

In summary, tool offsets and cutter compensations may be employed under the following circumstances:

a) Where a tool preset dimension is not exact, accounting for various tool holding methods (e.g. holders, collets, etc.).
b) To accommodate replacement tooling in cases where the originally specified tool is unavailable or has to be replaced because of damage.
c) Variations in tool size due to regrinds and refurbishment.
d) Changes in reference plane.
e) To accommodate slight variations in component size.
f) To accommodate roughing and finishing cuts using the same part program data.
g) To ensure accurate machining (of tapered and circular features) taking account of various tool tip radii.

7.4/3 Identification of offsets and compensations

It should be understood that both tool length offsets and cutter compensations are applied at the time of machining when actual cutter sizes are known. Generally, they are entered via MDI and stored within the CNC control unit. In some cases they can also be listed at the beginning of a part program tape or embedded within a part program itself. When the latter, any alteration via the console keyboard will override (replace) the previously loaded entry. The operator has access to both offset and compensation values at all times, and they may be entered, modified or deleted as required.

Both offsets and compensations are used by the control unit to make corrections to the programmed cutter path during machining. They have no permanent effect on the programmed data. Offsets and compensations are volatile in that they will be lost if power is removed from the control unit. Some control systems may allocate a specific offset/compensation value for use with a particular tool only; others may give such values an identity in their own right.

In just the same way that different tools are identified by a unique number (TØ1, TØ2, TØ6, etc.), tool offsets and cutter compensations may also be identified by a unique number. This latter number may or may not be the same identifier as the tool number. This makes it possible, although not necessarily advisable, to use different offset/compensation values with different tools.

When a different tool is required, within a part program, the miscellaneous function MØ6 is issued along with the tool function code and the requested tool number, e.g. TØ3. In addition, the tool change request may have to be accompanied by the associated tool offset or cutter compensation value to be loaded into the axis registers. The means of calling different offset values for different tools is specific to each individual control unit. It is common to include the offset register number in the tool function code, immediately following the tool number. For example, TØ3Ø2 calls tool number 3 with the offset values contained in offset register number 2.

> *Cutter compensation values can affect, quite dramatically, the programmed path of the cutting tool. This is especially important when such values are loaded, with the program, via punched tape. Tool offset and cutter compensation registers should always be checked before machining commences. Such values make no attempt to adjust programmed speed and feed values specified within the part program.*

Tooling probes
(Courtesy: Sandvik UK Ltd)

7.5 Proving part programs

7.5/0 Verification of part programs

Once a part program has been written it is imperative that it is verified as being safe to run. The safety of a part program relates mainly to the *programmed path of the cutter*. It is, arguably, fairly straightforward to code a part program to machine a particular component when working from a detail drawing. It is somewhat less straightforward to ensure that the cutter path does not foul projections that are likely to be present when the component is mounted on the machine. Clamping features are the most likely projections to be encountered. Even if due regard has been given to likely projections whilst programming, it is not uncommon for typographical errors to be present within the body of the part program. Imagine the consequences of accidentally omitting a decimal point from a dimension such as 11.23 mm.

The process of verifying part programs is known as **proving the program**.

It is part of the self-discipline of programmers, and CNC machine tool operators, to insist that all part programs are proven before being auto-run at the machine. Indeed it will probably be company policy to do so. The consequences of not proving part programs may range from damage to components or tooling, catastrophic damage to the CNC machine tool, or serious injury to the operator or other nearby personnel.

Additionally, just because a part program traverses a seemingly correct cutter path, it does not necessarily imply that the finished component will be correct in terms of dimension, geometry or surface finish. It will be necessary to verify that the component, as produced by the part program, is correct to drawing. It will also be necessary to verify that programmed feeds and speeds, as predicted by the part programmer, are applicable in the practical production situation.

It is common practice to have unproven punched tape in one particular colour. Once the part program has then been passed off as correct, a new different-coloured tape is produced to be used as the production tape. This system provides an immediate visual indication of the status of a punched tape, on the shop floor. Should an operator discover a program error, or a more efficient means of producing the part, then the part programmer concerned should actually make the correction. This is to ensure that any associated documentation can be updated and also that everyone is working to the current version of the part program concerned.

7.5/1 Visual inspection

This method represents checking, visually, the program resident within the memory of the CNC machine. Note that it is the ACTUAL part program to be run that is inspected and not the tape or the part programmer's manuscript. Factors to be checked include the programmed dimensions themselves, maximum movements in all programmable axes, cutter path when two axes are commanded to move simultaneously (especially under rapid traverse), depths of cut, changes in absolute/incremental dimensions, and the contents of offset and compensation registers.

This method represents the very least form of verification and should not be relied upon entirely. *All* part programs should be subject to such a visual inspection when installed in the memory of the machine tool control unit.

7.5/2 Single step execution

Following on from visual inspection and BEFORE auto-running, the part program shold be executed in **single step mode**. This means running the part program block by block. After a single block has been executed, the machine will stop. The next block will be executed when, and only when, the operator takes a conscious decision to proceed by pressing a button on the control console. All machines have such a facility and it should be utilised in all cases.

Single step execution should be carried out by utilising spindle speed and feed rate override facilities, provided on the control console. In this mode, feed rates are under the control of the operator and not the part program. This means that all axis movements can be monitored at a (slow) pace at which the operator feels most comfortable. Single step operation may be carried out with or without the component being mounted on the machine tool. When the component is not present, this technique is known as "cutting air". When the component is present, care must be taken to wind the cutting tool a safe distance from the work surface.

> *Proving part programs involving movement of the machine and/or the presence of the component creates a potentially dangerous situation. The operator must have no hesitation in hitting the emergency stop button if there is any doubt whatsoever about the anticipated motion of the machine tool. For this reason operators are urged to familiarise themselves with the location of the emergency stop button on the machine tool. Correct start-up procedures must be observed if this action is taken.*

7.5/3 Dry run

A **dry run** consists of running the part program in automatic mode. It is so called because there is no component installed in or on the machine, and "cutting" is done in air. The purpose of a dry run is to verify the programmed path of the tool under continuous operation. In doing so, the positions of any proposed clamping arrangements, or other projections within the set-up, can be observed for adequate clearance. A dry run will follow execution in single step mode. It will be carried out using the feed rate override facilities to slow down the speed of execution of the program.

7.5/4 Graphical simulation

It is becoming increasingly popular to apply **graphical simulation** techniques to prove part programs. A graphical simulation package emulates the machine

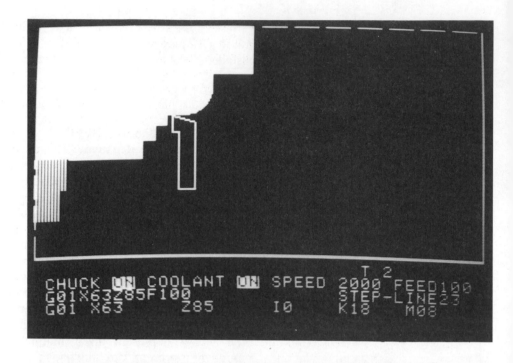

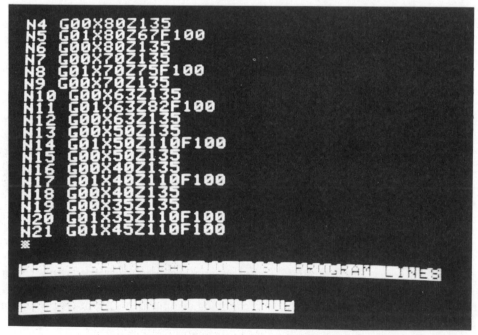

Fig. 7/19 A typical microcomputer-based graphical simulation system [*Aids Data Systems Ltd*]

tool and, using computer graphics, plots out the machine movement on a VDU screen. Machine movement often takes the form of a cutting tool shape moving around the screen according to the programmed machine movements. When the tool shape passes over a shaded representation of the component, it "erases" that part of the component. The resulting shape, left after execution is completed, represents the shape of the finished component. More importantly any gross deviations from the intended tool path can be observed and any potential crash situations highlighted.

With increases in memory capacity and the emergence of more sophisticated operating systems, many newer CNC machine tools now offer these features, built in within the control system software, as standard.

Alternatively, there are also a number of stand-alone microcomputer-based simulation packages designed specifically to carry out this function. Indeed many education and training establishments make extensive use of simulations. The benefits in the training situation are attractive. Students can be given hands-on programming experience in the mutual confidence that they cannot crash expensive machine tools. Many more students can be given hands-on programming experience since it is relatively inexpensive to set up a number of programming stations. There is no tool breakage or consumption of valuable raw material, and expensive machine tools are not subjected to abuse and misuse. Mistakes can easily be highlighted, quickly rectified, and the program re-run in a matter of minutes. Such a system is illustrated in Fig. 7/19. They are normally only suitable for components that can be produced from standard round (turning) or plate (milling) raw material, although within limits the dimensions may be user-defined. What is of some concern is that some packages adhere to programming standards that do not relate to the broad principles of full-sized CNC machine tools.

CNC turning computer simulation
(Courtesy: Aids Data Systems Ltd)

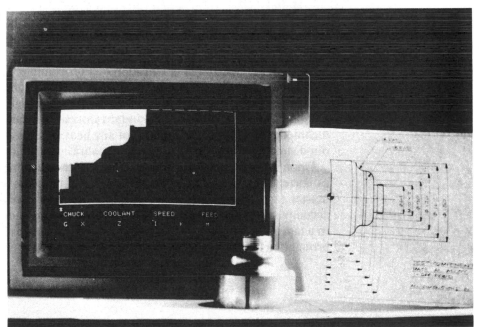

7.5/5   Pen plot

A relatively simple and inexpensive means of ascertaining the component profile is to substitute a pen tip for the cutting tool. In the case of a milling machine, this is a simple matter. A two-dimensional trace of the cutter path may then be reproduced, on paper, by placing a drawing board on top of the machine worktable.

On a lathe, some means of securing the pen to the cross-slide arrangement will be required. The resulting two-dimensional plot traces the path traversed by the tool during roughing and finishing cuts and also any approach and retract moves. It is easy to isolate any gross deviations from the intended cutter path.

7.5/6   First-off

The above methods of proving part programs relate mainly to the safety aspect of the machine tool movements. Whilst the finished shape of the component can be verified in broad terms, dimensional and geometrical accuracy cannot be assessed. A part program is not fully proven unless it machines a component to drawing. It follows that the final proving of a part program entails cutting an actual component. The component must then be subjected to dimensional, geometrical and physical inspection before the part program can be passed as correct. This is the **first-off component**.

Where the components are relatively small and made from readily available material, this merely involves using actual component blanks. A small amount of scrap can usually be tolerated. This is the ideal situation since it also allows programmed feeds and speeds to be "fine-tuned' to current machining conditions.

Where the components are large, or unduly complex, or already involve substantial oncosts from earlier operations, or are made from scarce or expensive materials, an alternative approach may need to be considered. In such cases, blanks of wood, polystyrene, wax or synthetic plastics are used as substitutes. It is imperative that any alternative materials should exhibit certain characteristics before being considered. For example, they must be capable of being machined without tearing, leave sharp edges, and resist any distortions imposed by applied cutting forces. They must be rigid enough to undergo handling, and inspection using standard workshop measuring instruments and techniques. They must be tolerant of any heat generated due to cutting, without distortion, and ideally should not be abrasive to machine tool elements.

Upon successful inspection of the first-off, the part program can be certified as correct. If amendments have been edited in at the machine, then a new master tape must be obtained. Ideally, and most conveniently, this can be accomplished by dumping the program from the memory of the control unit to a tape punch. Any previous tapes must be destroyed and all documentation brought up to date.

7.6 Computer-aided part programming

7.6/0 Manual part programming

The term **manual part programming** refers to the production of part programs directly in part program codes. It is still the most widely used technique for the production of part programs. To accomplish manual part programming successfully, specialist skills and knowledge are required by the part programmer. These include:

a) The ability to read component drawings.
b) A sound knowledge of trigonometry, geometry and mathematical calculations.
c) An understanding of workshop practice, machine tool operation and cutting tools.
d) A working knowledge of materials, feeds and speeds.
e) An understanding of workpiece location, clamping and fixturing.
f) A knowledge of part program terminology and codes (possibly including techniques of data input).
g) A thorough knowledge of the capabilities and facilities offered by the CNC machine tool.
h) The ability to think clearly and logically with attention to detail.

Manual part programming is thus labour-intensive needing skilled personnel. The time involved in producing part programs is proportional to the complexity of the component. For complex three- and four-axis machining, this can involve many man hours of work. The situation is aggravated if the programmer needs to have a knowledge of more than one CNC machine tool or control system. Once the part program has been written, it must be converted into a machine-usable form, usually via MDI or punched tape. It must be verified and proven, tying up more man and machine hours.

It will be appreciated that manual programming can be very time-consuming and somewhat inefficient. The manual involvement at every stage presents the added possibility of incurring errors. The costs involved in producing the part program must be borne by the components eventually produced and can thus amount to a substantial proportion of the total cost. The time required to produce the part program can also form a significant part of the total lead time in the production of finished components. In some cases (machining in 5 or more axes), it may be impossible to produce part programs by manual programming methods.

7.6/1 Concept of computer aided part programming

The concept of **computer aided part programming** (CAPP) is to enable a computer to generate the part program code required to finish-machine the component. Advantages of achieving this include:

a) Part programming procedures will be considerably simplified.
b) The production of part programs will be speeded up.
c) Computations of cutter path coordinates will be transferred from the programmer to the computer, thus reducing the possibility of errors.

d) The part program will be generated immediately into efficient and accurate machine-usable code, again alleviating the possibility of errors.

e) The part programmer only has to learn a single (English-like) language, no matter how many different machine tools or control systems might be involved.

f) Such systems can deal with many axes of simultaneous movement.

g) The generated part program code can be transmitted to the machine tool, automatically, via DNC if required.

Although CAPP languages differ slightly, there are normally three stages in the process:

1) Component Shape Definition
2) Processing
3) Post-processing

7.6/2 Component shape definition

From the component drawing the programmer defines the required shape in terms of descriptive geometry elements using abbreviated English-like terms. The component shape will be split up, and described in terms of points, lines, complete circles, distances, and directions. The above elements are assigned consecutive numbers to identify their position and differentiate one element from another. Simple examples might be:

```
P1 12.35 35.5     -   Point No.1 is at co-ordinate position X12.35,Y35.5
C2 3.5 3.0 1.7    -   Circle  No.2  has  a radius of 3.5 with the centre
                      positioned at co-ordinate X3.0,Y1.7
W5 P12 P13        -   Distance  W5  is  the straight line distance between
                      Points P12 and P13
```

There is often facility for including arithmetic operators and functions within the geometry definition. For example, the following definition would be quite valid:

```
P11 SIN(39.5) 23.2*6.2  -  Point No.11 is  at  the  co-ordinate position
                           specified by X=SIN(39.5) and Y=(23.2 X 6.2)
```

Depending on the system, graphics capability may allow illustration of the geometrical construction on VDU terminals. In such cases the system is known as **Graphical Numerical Control** (GNC). Such systems will also have added facilities, for example the ability to show the programmed tool path dynamically on the graphics screen. The time to cut the part can also be generated from the programmed speed and feed data. There may also be facility for outputting the component shape, and the programmed cutter path, to other peripheral devices such as graphics printers or plotters.

All input definitions will be accepted (whether correct or not) and written to an input file. Comprehensive interactive editing facilities will be available to amend geometry definitions.

7.6/3 Processing

When shape definition is complete, the input file will be processed. **Processing** generally comprises two parts. Firstly, the input file is interrogated and checked for errors. If errors are found then an error file is generated. The error file identifies the source of the errors and alerts the programmer using comprehensive error messages. The input file must be edited and corrected before the second stage of processing proceeds.

When the input file is valid, additional information regarding tool sizes, machining sequence, etc. is input. Processing then converts the geometry definition into all the data required to machine the component, excluding considerations applying to the machine tool itself. This data forms the output file. Usually this comprises the dimensional description of the cutter path from a specified point. Time taken to machine the component based on current feed and speed values may also be provided. This information is "raw" positional information and is known as **cutter location data** or CL Data. Note at this point that the information does not contain G-codes, M-codes, etc., it is completely machine tool independent. Thus, the CAPP system can be used up to this point regardless of machine tool. All part programs may be produced with a knowledge of only one system.

7.6/4 Post-processing

A **post-processor** or **link** is a computer program that translates the output file and assembles it into a form which a particular machine tool can use. This means adding in G-codes, M-codes and other machine-specific information, in the correct format. Since different controllers interpret identical codes for different purposes, and require the same information but in different formats, each machine tool requires a unique post-processor. However, post-processors are relatively simple computer programs and can be provided at a relatively small cost.

This approach makes a computer aided part programming system extremely flexible.

a) One system can support many different machine tools.

b) Only a single "language" needs to be learnt to program a variety of machine types.

c) If existing machines are updated to offer enhanced facilities, a simple update to the post-processor software will allow such facilities to be utilised immediately.

d) If new machines are acquired, only a relatively inexpensive post-processor need be purchased to integrate them into the existing, familiar system.

e) Other CNC-based equipment in other application areas can be accommodated. These include punching, grinding, EDM, etc.

7.6/5 Computer aided part programming systems

The introduction of a computer aided part programming system means additional capital expenditure. Almost by definition it implies that programming shall be a separate and distinct function requiring support. There are, however, a number of ways in which such systems can be implemented.

a) **Stand-alone systems**

Many systems can be run on self-standing mini- or micro-computer systems. This means that the system can be installed and maintained on site. The system can be used for additional tasks within the organisation. Access and security of data will be much simplified. This option will probably incur a heavy initial capital investment and ongoing hardware and software maintenance costs.

b) **Time-share rental**

A second option is that of time-share rental. Access to a mainframe-based system via the telephone network may offer a cheaper alternative without committing the organisation to large capital expenditure. Running costs are likely to be higher since telephone charges will be additional to rental time on the system. Batch run operation may slow down response time, and security of information may be a cause of some concern. The user will need to learn the part programming system but will probably be able to draw easily on support services. The user will be free from any maintenance or update obligations on either hardware or software.

c) **Bureau**

It is also possible to contract out part programming to a specialist computer aided part programming bureau. Whilst this may be expensive on a job-to-job basis, no real commitment or large capital outlay is required. Turnround times may be unpredictable and security of information may be of some concern. No commitment to training, maintenance or updating costs will be needed.

The decision to employ CAPP techniques and the choice of operation are management policy decisions that can only be taken in the light of current and future business considerations. With the latter two options above, for example, there is no scope for using the same computer system to form an integrated database from which information can be extracted for other uses. Similarly, the introduction of DNC or flexible manufacturing systems could not easily be based on such an arrangement.

1. What is a part program and why is it so called?
2. Define the terms "command", "format", "word address", "block", and "character" in the context of part programming.
3. Part programs must conform to a programming format. Explain the difference between fixed block, variable block, and word address formats.
4. Explain the terms "preparatory function" and "miscellaneous function", stating where they would be used in a part program.
5. Explain the functions of the following M-codes: M$\emptyset$2, M$\emptyset$6, M$\emptyset\emptyset$, M$\emptyset$5 and M$\emptyset$3.
6. Explain the functions of the following G-codes: G$\emptyset\emptyset$, G$\emptyset$1, G71, G9$\emptyset$, G8$\emptyset$ and G$\emptyset$2.
7. State, giving examples, what is meant by a modal function.
8. Write the part program code that would typically start a part program. Give a line-by-line explanation of each block.
9. What precautions should be taken if a CNC machine tool is stopped, in mid-program, by means of the emergency stop button?
10. State the differences in part program code between programming for both linear and circular interpolation.
11. What is a canned cycle? State typical canned cycles for both Turning and Machining centres.
12. Define, outlining the differences between, a loop, a macro, and a sub-routine. Give examples of where each technique would be used.
13. Explain, giving examples, the features of reflection, rotation, translation, and scaling as extended part programming facilities.
14. What are tool length offsets; when are they used and how are they programmed?
15. What is cutter compensation; when is it used and how is it programmed?
16. Explain the importance of proving part programs and state *four* methods of doing so.
17. Compare and contrast manual programming and computer assisted part programming, stating any relative advantages and limitations.
18. What is conversational part programming and where would it be employed?
19. What is a post-processor and where would it be used?
20. Describe the events that should be considered before deciding to part program a particular component.

Associated considerations 8

8.1 Flexible manufacturing systems (FMS)

8.1/0 The organisation of production

When a manufacturing organisation decides to produce a component, it has to decide on the best way of organising its resources. The main resources will be people and machines. Machines will have to be physically arranged to produce the components in the quickest, most efficient and cheapest way possible. The better this is achieved, the more competitive and more profitable the organisation will become. In addition, the workforce will need to be managed effectively in order to realise the full potential of the physical resources employed. People are perhaps the most useful resource in that they can think, apply judgement in unforeseen circumstances, adapt to new situations, and are capable of learning and improving on the skills they possess. People are, however, expensive to employ, they require rest and services, can be unpredictable and prone to making mistakes.

The requirements of an ideal production system include:

- Low operating costs.
- Consistent quality.
- High machine utilisation.
- Low setting-up and job changeover times.
- Low work in progress.
- Minimum delay between receipt and delivery of orders.
- Predictable manufacturing times.

A **Flexible Manufacturing System** (FMS) is, first and foremost, a *way of organising for production*.

Before the concept of FMS is discussed it will be helpful to examine the conventional approaches that have been followed in organising for production. Traditionally, for multi-component production, three approaches can be identified:

1 PRODUCT (or Flow) Layout
In this approach the plant and equipment are laid out according to the requirements of the product. Machines are arranged sequentially to carry out

operations on the product in a strict order. This is typical of *flow production* and gives rise to the term *flow*, or *production line*. Each production line is dedicated to a specific component (or component range). Although many production lines may be present, components cannot be transferred, or interchanged, from one line to another. This type of production is normally associated with the mass production of motor cars.

Disadvantages include:

High volume is essential for high machine utilisation.

Almost impossible to accommodate different products.

Breakdown of a single machine can immobilise production.

Can require high levels of (semi-skilled) personnel.

2   PROCESS (or Functional) Layout

In this type of layout, plant and machinery are grouped according to function. Thus, all lathes are grouped together within a "lathe section", all drilling machines within a "drilling section," and so on. To ensure the fullest machine utilisation, components are routed to the most currently available machine. Frequently, however, batches of components often form queues whilst waiting to be processed. They may also re-visit a process already visited earlier in the production cycle. This means that there is a high level of *work-in-progress* (WIP) and material handling. Work in progress refers to part-finished components waiting, or queueing, for subsequent operations. Since there is often a variety of components being processed at the same time, throughput times vary, dependent on the current level of WIP.

Disadvantages include:

Complicated routes for components make control difficult.

Machine loading is erratic.

Queues can cause high investment in raw material and WIP.

High proportion of setting and re-setting time.

Throughput times are difficult to predict.

Skilled personnel required.

3   GROUP TECHNOLOGY Layout

Group Technology is an arrangement of machines, tools and services to facilitate the complete manufacture of a defined range of similar products. Machine groups comprising different numbers of different machines form what are known as *machining cells*. Each cell is able to produce any component from within a carefully pre-defined set of similar components. Sets of similar components are known as *component families*. The layout of individual machines within the cells is for convenience. Generally, the number of operatives within the cell is less than the number of machines.

Disadvantages include:

Machine under-utilisation is inevitable.

Unworkable unless suitable component families are identified.

Multi-skilled personnel are required.

The three types of production are illustrated in Fig. 8/1.

In each of the above cases, the disadvantages outlined conflict with the requirements of the ideal production system. The concept of flexible manufac-

Fig. 8/1 Three traditional methods of organising for production

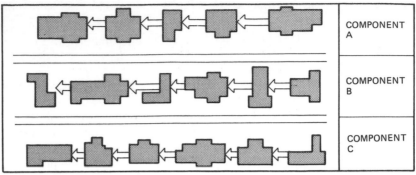

(a) **FLOW PRODUCTION**

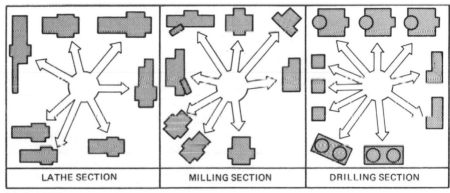

(b) **FUNCTIONAL LAYOUT**

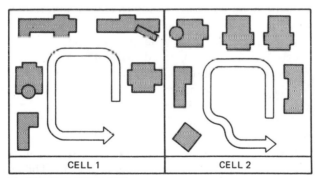

(c) **GROUP TECHNOLOGY**

turing is designed to eliminate many of the above disadvantages and dramatically cut the overhead costs of manufacture by introducing high levels of automation under computer control. The term *overhead* means any cost that has to be made which is not directly attributable to component manufacture. Overheads include the amount of money tied up in WIP, stockholding and raw material, the costs of employing indirect labour (inspectors, tool/machine setters, progress chasers, etc.), and support services (such as canteen facilities, safety and first aid facilities, office staff, etc.) for the workforce.

8.1/1 Flexible manufacturing

Flexible Manufacturing Systems embody four main principles:

1 Unmanned Operation

Flexible manufacturing systems are usually capable of unmanned, continuous operation for at least one shift. This can dramatically increase machine utilisation and productivity. Traditional 8-hour shifts (which have evolved as the period of time a man can reasonably be expected to work at one stretch) need no longer be boundaries within which production schedules must fit. Smoother production schedules may be planned and people need not be required to work at unsocial times during shift working. Unmanned operation will dramatically reduce labour costs.

Unmanned operation however, implies the utilisation of very sophisticated support facilities. Different part programs must be identified and downloaded (via DNC systems) to different machine tools automatically. Components need to be loaded, unloaded and transported automatically. Cutting tools need to be monitored for wear or breakage and new tools automatically loaded as appropriate. Swarf needs to be cleared from the machining area and disposed of automatically. Automatic washing and inspection facilities may also have to be provided.

Such support services are expensive to implement and, along with the high costs of CNC machine tools themselves, account for the high capital investment required in setting up true FMS installations.

2 Random Component Production

True FMS allows the random launching of components into the system. This is made possible by the ability to call up different part programs at different CNC machines almost immediately, with the ability to automatically select, transport and load components. Queuing, work-in-progress and large stock levels are largely eliminated. Machines are kept fully utilised. Throughput times can be accurately predicted and short lead times can be maintained. Machine breakdowns can be accommodated by re-routing components to other machines. No longer are large production runs required to justify high capital investment and long setting times. Indeed, FMS can be competitive with batch sizes as low as one.

The ability to launch components into the manufacturing system at random is one of the most important benefits of FMS. The term FLEXIBLE manufacturing does not necessarilly mean flexible enough to produce a large variety of components, but flexible enough to *produce components as and when they are required*.

3 Automatic Tool and Component Movement

FMS cannot be properly implemented without the ability to automatically select, transport, and change cutting tools and components.

Many CNC machine tools have integral tool magazines comprising up to 160 tools that can be changed automatically. Automatic production of a variety of components, especially during unmanned working, may exceed the capacity of integral tool magazines. It may be necessary to arrange that the tool magazines themselves are replenished automatically. This can be achieved by detachable tool magazines that can be loaded by robots. The situation is further complicated by the need to provide replacements for worn or damaged tooling. This is normally known as "sister tool replacement".

Components have to be delivered to the machines, loaded and unloaded and, when finished, transported away to other machining operations, washing/inspection stations, or storage locations. Robots and/or automatically guided vehicle systems are a common element of most FMS installations, although fixed conveyor systems are also employed.

4 Stand-alone Operation

In its simplest form an FMS will comprise at least two CNC machine tools, probably with automatic tool changers and serviced by a single robot for component handling. In such cases it is more likely to be termed a **Flexible Machining Cell** (FMC). At the other end of the scale it may be an extensive and comprehensive system integrating all aspects of manufacture quite automatically. It is an advantage that an FMS can start life as a small machining cell and be built up, modular fashion, into a fully integrated manufacturing facility.

It is vital that all the machine tools making up an FMS retain the ability to work as stand-alone machine tools and retain their individual flexibility. This ensures that production is never disabled completely. It is common for FMS systems costing millions of pounds to take between 2 and 5 years to install. The layout of an FMS system is illustrated in Fig. 8/5.

The key to successful FMS is **software**. The technology of CNC machine tools, DNC, robots, material handling, adaptive control, etc. have been successfully implemented, in isolation, for a number of years. The key to bringing them all together to work in harmony as an FMS is computer control, and this means software. It is not uncommon for up to 25 man years of software preparation to be invested in large FMS installations. This is a third reason why FMS systems are expensive to install. It can also be argued that it is easier to alter the software should the manufacturing requirements change.

It is easy to view FMS as the sole province of metal machining. This is not the case. Flexible manufacturing is a way of organising for production and as such can cut across all manufacturing disciplines. Indeed, true FMS should incorporate automated storage and retrieval of components, automated inspection, assembly and test. In practice, this is very rarely realised. In the context of this book we could be justified in interpreting the term FMS to mean Flexible Machining System, although this term does prove somewhat restrictive.

8.1/2 Component families

It would be wishful thinking indeed to suppose that we could design an FMS which would cope with the production of every conceivable component imaginable. How, then, is an FMS designed in the first place?

It was stated earlier that both Group Technology and FMS can only be implemented successfully if suitable **component families** can be identified. The first step to implementing FMS is a survey of the components being produced.

Similar components are collected into "families" exhibiting broadly similar characteristics. Once a family has been identified, a "composite component" may be envisaged. A **composite component** is one which contains all the features of all the members of the family, although it may not actually exist physically. Components produced predominantly by turning operations are known as *cylindrical* or *turned* components, and components not predominantly cylindrical in shape are known as *prismatic* components. An example of a typical component family and the resulting composite component are shown in Fig. 8/2.

Fig. 8/2 Examples of a typical component family and its composite component

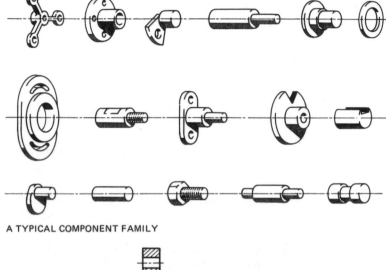

A TYPICAL COMPONENT FAMILY

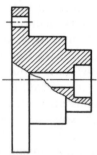

THE COMPOSITE COMPONENT DERIVED FROM THE FAMILY

In a small organisation the components that will eventually make up a family may be identified by eye. This is rather haphazard, and difficult when the number of components becomes large. A coding or classification system is normally used. Several such systems exist but basically each component is examined and given a code number comprising between 6 and 10 digits. The digits will be applied according to such items as:

Function and structural shape.
Shape of raw material required.
Processing operations required.
Material type.
Physical size.
Accuracy required.
etc.

Once each component has been allocated its corresponding classification, a computer can scan the code numbers and sort them into naturally occurring families. Families often emerge as sets of nearly identical code numbers.

Production data is then collected for each component in the family, for example:

Machining operations and sequence.
Tooling and workholding requirements.
Quantities and frequency of demand.
etc.

This data is then analysed to determine the types and numbers of machines required, capable of machining all components within the family. These machines will then form the building blocks of the FMS.

8.1/3 Automated work transport

Essential to the operation of FMS is the automatic selection, delivery and collection of components. This can operate at a number of levels.

The first level concerns *getting the components into and out of the machine tools*. The most common solution is a robot or robots loading and unloading components between machine tool and conveyors. Where the components are relatively small and manageable, this represents a convenient solution. Where components become large, it is usual to employ some form of **automatic pallet changer** (APC). A pallet, in this case, is a sub-table onto which the component is mounted, and remains so, during machining. There is usually more than one pallet station on each machine. Twin or multi-station APCs are available. Whilst machining is taking place utilising one pallet, the other is being loaded with the next component. The pallets are automatically switched at the end of the machining cycle to allow the removal of the finished component (and loading of the following component), whilst machining carries on uninterrupted. Typical APC configurations are shown in Fig. 8/3.

At a second level, components have to be *transported between various machines and other stations* within the system. This can be achieved by continuously running or command-driven conveyor systems. This is probably the least expensive, easy to install option. It is also the least flexible if changes have to be made.

An alternative arrangement is to use **automatically guided vehicles** (AGV). An AGV is a driverless seemingly free-ranging vehicle that can be commanded to take different routes, and load or discharge its component cargo, under computer control. AGVs are more flexible than fixed conveyor systems in that their routes can be easily altered and they can accommodate a considerable variety in the size and number of components they can carry. They can also be used for other tasks such as transporting tooling or swarf cargos.

AGVs can be rail guided, guided by laser, guided by light-utilising photoelectric sensors, or more commonly by inductive wires concealed underground. The latter system is relatively simple and cheap to install, is flexible, and does not create fixed obstacles on the shop floor. It can easily be added to by simply cutting additional grooves in the floor and connecting additional guide wires. Wire-guided systems can negotiate awkward routes since they can turn corners easily. Each AGV must, however, provide its own motive

Fig. 8/3 Typical automatic pallet changer configurations

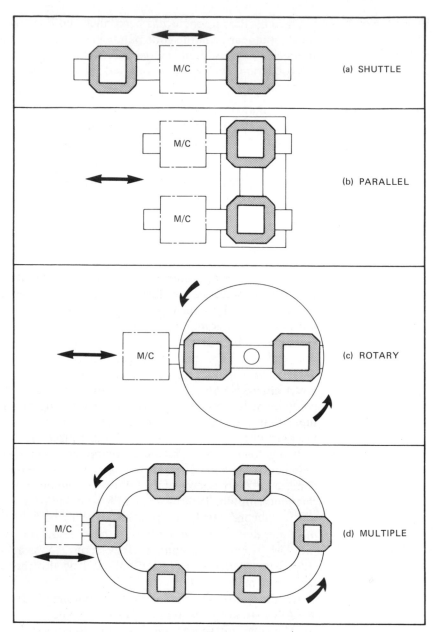

(a) SHUTTLE

(b) PARALLEL

(c) ROTARY

(d) MULTIPLE

power. The inductive wires provide route guidance only. Invariably the vehicles are electrically driven by their own on-board batteries. These batteries can then be re-charged at convenient times during the manufacturing cycle.

Each AGV must house a range of sensors to convey information about its current status: for example, positional information, the presence or absence of a cargo, collision sensors, docking sensors, and so on. In order to reduce the complexity of the software required, it is common for AGVs to house their own on-board computer controllers which can be individually route-programmed. The cargo must be identified by the system in some way. Common methods include mechanical or optical coding systems based on the binary system.

Fig. 8/4 Representation of an automatic store and retrieve warehouse

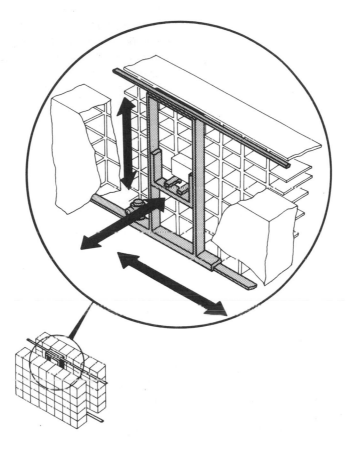

Fig. 8/5 The layout of an FMS system

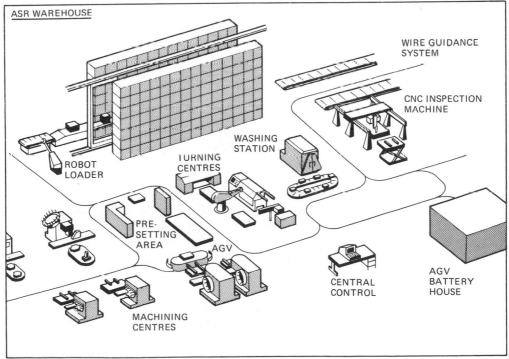

At a third level components may be *automatically stored*. In a complete FMS this would be achieved by an **automatic storage and retrieval** (ASR) warehouse. Components are housed in standardised containers which themselves are housed in a 3-dimensional matrix of storage racks. A dedicated machine, resembling a cross between an overhead crane and a fork lift truck, identifies the cargo and automatically stores or retrieves the container in a predetermined storage position. A typical ASR warehouse is illustrated in Fig. 8/4.

ASR warehouses can provide accurate control on the movement of stock. They are very space efficient and offer greater security and reduced damage and loss due to material handling. They are, however, extremely expensive to implement and may provide problems if the system breaks down.

An FMS will include, to a greater or lesser extent, the above features. An example of an FMS layout is shown in Fig. 8/5 and the photograph on page 38.

8.2 CAD/CAM

8.2/0 What is CAD/CAM?

The term CAD/CAM stands for Computer Aided Design and Computer Aided Manufacture, and is intended to relate to a closely integrated system of producing components from the design stage through to manufacture using computer assistance. In practice it is only larger installations that achieve this goal. It is more common to find that CAD systems (based on computer aided draughting principles), and CAM systems (based on stand-alone CNC machines), operate in comparative isolation. True CAD/CAM systems provide the all-important link that converts information generated by the CAD system into a form that can be readily utilised by the manufacturing systems of the organisation.

8.2/1 Computer aided design

CAD systems require large amounts of computing power. For this reason they are usually based around mini-computers or large-capacity microcomputers. Peripheral devices such as printers, plotters, disc storage devices, etc. are also required. These may be shared by a number of users. CAD software is usually expensive.

It is common to talk of a **CAD workstation** at which the designer interacts with the system. A typical workstation comprises a high-resolution graphics screen and some means of inputting information. Input devices include keyboards, graphics/menu tablets, light pens, joysticks and hand-held "mice". On larger systems it is common to have an additional VDU screen on which commands, menus and other textual information appears, all graphics being plotted on a dedicated graphics VDU. The designer using a CAD system now spends most of the working day sitting at the computerised workstation and concentrating on information displayed on VDU devices. Many of the above input devices have been designed to ease the strain of interacting with computerised systems. These and other ergonomic considerations, such as lighting and posture, should be considered as an important aspect in the planning of a CAD installation.

CAD software is usually supplied as different *packages* that can be used for different aspects of the design process. A CAD system is likely to consist of more than one package. It is essential that all packages within the system are compatible so that data can be transferred between them. The most common CAD packages include:

Design Analysis Packages

This is specialist software that performs complex mathematical analyses on designs originated by the designer. Stress analysis on large structures is a typical example. Such packages have the advantage that designs can be optimised for best performance in a very short time. Better quality products with fewer errors can be produced quickly and accurately.

Surface and Solid Modelling Packages

This software allows the creation of 3-dimensional models of objects to be constructed quickly on a graphics screen. Design concepts can be visualised from all angles, in true perspective and, in some cases, colour. Surface modelling generally produces wire-frame representations, whilst solid modelling produces a solid representation which can be sectioned as required. Modifications to the design can be carried out swiftly allowing various alternatives to be explored. Customers' requirements can be matched quickly and accurately and design time is considerably shortened.

Simulation Packages

This software usually models the operation of dynamic systems. For example, mechanical systems such as car suspension performance, or organisational systems such as complete factory layouts simulating different component routing through the system, may be simulated. Performance characteristics and design variables can be optimised in a very short time without the need to construct physical set-ups.

2-D and 3-D Draughting Packages

These are by far the most common packages in use. They allow detail drawings and designs to be constructed quickly and easily on the graphics screen. Complete working drawings can then be generated from the stored information. In addition to the facilities of

Move, draw and delete lines;
Points, circles, arcs, etc. in a variety of line types;
Ability to store and retrieve information to/from backing store;

common facilities of such systems also include the following:

Rotate all or part of the design through any angle.
Scale parts of the design up or down.
Translate parts of the design to different positions.
Replicate the same feature(s) at different places.
Pan across a large design using the screen as a window.
Zoom into or away from design features.

In addition many packages offer automatic dimensioning and cross-hatching facilities and the ability to apply layering. *Layering* involves producing different parts of the design on different screens and storing them separately. The designer can then recall any screen combination and superimpose different layers on top of each other. This allows an uncluttered build-up of drawings to be accomplished.

Standard parts or symbols may also be defined and stored in a *symbol library*. They may then be recalled and positioned on the drawing at any desired position.

The data generated by the software is held on disc or tape as computer files. If this data can be accessed by other software within the manufacturing system, it can then form the basis of a manufacturing database that can support post-design activities.

8.2/2 Computer aided manufacture

The most readily understood form of CAM is the stand-alone CNC machine tool, closely followed by DNC operation. The definition of CAM, however, should be understood in a wider context. Non-metal-cutting production processes (welding, presswork, EDM, etc.), and related activities such as computer aided inspection, testing and assembly techniques, together with automated materials handling, should be included.

CAM software is perhaps just as important. Much CAM software makes use of data generated at the CAD stage. By interrogating the appropriate computer files, CAM software can provide:

a) Bill of Material and parts list generation which can be used for materials requirements planning (MRP), stock control and estimating.
b) Post-processed output of component design data into CNC part program language suitable for transmission via DNC.
c) Production planning data concerning machining times, operation sequences, and tooling requirements which can assist process and production planning.

Much of the above CAM software is itself now being packaged into a separate category known as **Computer Aided Production Management** (CAPM). The aim of a CAPM system is to monitor and control the production management elements of materials control, work-in-progress, and accounting.

It can be seen that a manufacturing installation making use of CAD/CAM, FMS and CAPM facilities becomes a totally integrated manufacturing system. Benefits come from each sector of the manufacturing facility working on the same, current and up-to-date information from a system database. Such information is available immediately and much of the routine and time-consuming clerical work is dealt with by computer. The majority of paperwork, documentation and records are also computer-generated.

Rarely do such systems become established by design. They are usually formed by introducing computer-based facilities into existing set-ups. The benefits derived from installing individual systems cannot be equated to the full potential of a truly integrated system. Reasons for this include: missing links in the manufacturing chain; already established work practices, procedures and systems; traditional departmental boundaries; resistance to change, etc.

The modern production engineering philosophy advocates a total **Computer Integrated Manufacturing** (CIM) approach as a total design concept. Such approaches are rare since the amount of preparatory work in design and planning, and the huge capital investment required in plant and machine tools, are often prohibitive.

8.3 Adaptive control

8.3/0 The need for adaptive control

Adaptive control refers to the *in-process sensing* of tool wear, component size, and component quality, in order to provide automatic compensation of machining process parameters to maintain consistent accuracy and quality of output. In-process sensing means that sensing (and subsequent compensation) is carried out whilst actual machining is taking place.

The most important component in a CNC machine tool, and indeed an FMS, is the tool tip. If the tool tip breaks down then the whole system is rendered ineffective and the costs of production increase.

a) Tool wear can result in oversize components being produced, necessitating rework operations.
b) Tool wear can reduce optimum cutting conditions, resulting in components of inferior quality (and further tool wear!).
c) Tool wear can cause premature tool breakage.
d) Tool breakage can scrap expensive components, damage the machine tool, and cause serious injuries to operatives.
e) Tool breakage can disrupt machining and production schedules.

Adaptive control implies two things. Firstly, that the tool and component characteristics of size and quality can be sensed whilst machining is taking place and, secondly, that there are process parameters which can be automatically adjusted in a responsive manner. In conventional machining situations, the operator provides the adaptive control. The experienced machinist is able to assess the quality of the cutting tool by sight, sound, touch and smell, and can then effect control by adjusting the speed, feed, depth of cut, etc., or by replacing the tool. In automatic adaptive control, speeds, feeds and depths of cut are still the process variables which are controlled. Monitoring of tool and component condition, however, has to be done by some form of sensing.

8.3/1 Tool wear sensing

Tool wear can be predicted using well-tried tool life calculations. A tool may thus be replaced after x-minutes cutting or when y-components have been machined. This method is unsuitable in the modern manufacturing environment for the following reasons:

a) Certain (often incorrect) assumptions have to be made regarding the formulae used.
b) Ideal cutting conditions are assumed in the calculations.
c) New tool materials are constantly being introduced.
d) Difficult-to-machine workpiece materials are often not accommodated.
e) The method is based on "trial-and-error" formulae.
f) A more reliable method is required.

The majority of adaptive control systems sense the onset of **tool wear**. This may be done in a variety of ways. In all cases an effective system must

Properly reflect the degree of wear at the tool tip.
Have quick response.
Be sensitive to sudden changes in cutting conditions.
Be robust and adaptable for use on the machine tool.
Be reliable and require little maintenance.
Be available at a realistic cost.

Sensing methods may be contact or non-contact. They may also be either *direct*, measuring dimensional, volumetric or other changes in the tool itself, or *indirect* via other conditions that can be more conveniently measured. Three classes of adaptive control are described below.

1 Feedback from the Tool

Electrical sensors monitor the change in electrical resistance across the tool/workpiece junction when the tool is in contact with the workpiece during cutting. As the tool wears, its contact area with the workpiece increases and the electrical resistance across the junction decreases.

Pneumatic sensors monitor air back-pressure formed between a nozzle set in the clearance face of the cutting tool and the workpiece surface. As the tool wears, the gap between the clearance face of the tool and the workpiece surface decreases resulting in an increase in back-pressure.

Optical sensors illuminate the wear zone of the tool tip when it is not in contact with the workpiece. The light beam is reflected to impinge on a photo-transistor. The output signal from the photo-transistor can be correlated to tool wear. This system can only be used when the cutting tool is not continuously in contact with the workpiece; for example in milling, when the cutting edge can be exposed to a light beam during that part of its rotation when it is not in contact with the workpiece.

2 Feedback from the Workpiece

Contacting or non-contacting transducers measure and compare actual workpiece dimensions with commanded workpiece dimensions. Tool wear is directly related to dimensional changes in the component. Inspection (touch trigger) probes held in the tool spindle are a common way of implementing this approach.

Fibre optic sensors probe the same path as the cutting tool and measure the reflectivity of the newly machined surface. Reflectivity of the surface varies with the roughness of the surface texture. The deterioration in the quality of the surface of the workpiece is taken as an indication of tool wear. Research has also been carried out on this theme using laser systems.

Electronic feeler displacement transducers measure the distance between the workpiece and the toolpost. The output signal of the transducer varies as the displacement varies and this is proportional to tool wear.

3 *Feedback from the Cutting Process*

Thermocouple transducers measure the cutting temperature at the tool/workpiece junction. Cutting temperature will increase as tool wear increases. Infrared sensors can also be focused onto the cutting zone to measure the cutting temperature.

Piezo-electric transducers located near the cutting tool edge respond to changes in vibration. An increase in vibration indicates the probability of the onset of tool wear.

Power and torque transducers (wattmeters and accelerometers respectively) mounted within the machine spindle detect increase in power and torque respectively. As the tool wears, rubbing takes place and the power and torque values increase. These values need to be preset prior to machining to provide reference levels. This will probably be accomplished by cutting a one-off with new, sharp tools and recording torque and power levels. The programmer then has to decide what increase in level constitutes a worn tool.

Tool wear sensing is important since tool wear is probably the biggest cause of tool breakage.

The control system software will need to be programmed to take appropriate action conditional on the output from the adaptive control system. This may be to cease operation and provide an alarm system requesting attention from an operator, or automatically initiate a tool change for a sister tool. If the sister tool replacement approach is adopted, then sister tools obviously have to be provided in the tool magazine. This means either duplicating all tools needed to produce the component or predicting which tools are likely to wear or break and hence require replacement.

8.3/2 Broken tool detection

A complementary facility to an adaptive control system is that of **broken tool detection**. It is normally provided as an emergency check rather than a routine since it takes valuable time to implement. It may be accomplished in a number of ways.

One method is to provide a light spring-loaded switch mounted to the side of the machine. At various times during the machining cycle, the tool is traversed to negotiate this switch. If no signal is received by the control system, it is assumed that the tool is either missing or broken

A second method involves passing the tool through a caliper to break either a light beam or an air jet. With intelligent application this method can also be used to automatically set the tool length offset of a replacement tool.

If a broken tool is detected, the course of action to be taken is a little more complicated. Obviously the broken tool needs replacing, but other questions need to be addressed:

Is the broken tool still in the workpiece?
Should the component be rejected and the next one loaded?
Should machining continue but at a later point in the program?
Should the part be probed to identify the presence of a broken tool?

Adaptive control is probably the major missing link in a totally automated machining system. It must be considered an extremely important aspect of unmanned machining installations.

8.4 Industrial robots

8.4/0 What is an industrial robot?

Robots are increasingly becoming a key feature of modern industrial manufacturing systems. But what is a robot and how does it differ from traditional "automation"? Automation can be described as the capability to operate without direct human intervention. Automatic devices have been on the industrial scene for more than 100 years. They are largely mechanical devices purpose-designed and built to perform a dedicated task. Because the configuration cannot easily be changed, a modern term labels this as *hard automation*. If the task changes, then this "hard" automation becomes redundant, or has to undergo physical modification to adapt its operation to suit the new task, or may be cannibalised and the components used again on other applications.

Industrial Robots also operate without direct human intervention and as such form a sub-class of automation. In addition, they exhibit certain other characteristics that set them apart from dedicated automatic devices.

a) **Industrial robots are programmable**   The movements and actions of an industrial robot can be determined initially by a human operator. The sequence of movements can be stored within a control system computer and the robot commanded to perform these movements repeatedly and with great accuracy. If the task changes, the robot can be "taught" a new sequence of movements thus easily adapting to the new task. Since operation is largely under software control (i.e. the control of a computer program), robots can be thought of as "soft" automation.

b) **Industrial robots have multiple movements**   Industrial robots can perform 3-dimensional manoeuvres in space. Six degrees of freedom of manipulation are provided on most industrial robots, which make them adaptable to a wide variety of industrial tasks. Additional degrees of freedom may be provided to increase the accessibility of the robot for certain applications. The 3-dimensional space within which the robot can reach is known as its *working envelope*.

c) **Industrial robots have interchangeable grippers**   Most industrial tasks involve transporting or manipulating tools, components, appliances or applicators to perform various manufacturing activities. The majority of these tasks involve the holding or gripping of items of varying size and shape. Since it is almost impossible to design a universal holding device, and in order to retain the flexibility of robot applications, most robots can be fitted with a variety of "hands". These are known as *end effectors* and will often be purpose-designed for a particular task. They may be mechanical, electro-magnetic, or pneumatic (vacuum cup) devices or consist simply of hooks or holders. It is becoming more common for robots to be applied to more than one task simultaneously, thus necessitating an end effector change from one task to another. This is similar to the automatic tool changing facility available on many CNC machine tools.

On the basis of the above characteristics an industrial robot may be defined as:

a computer-controlled, re-programmable mechanical manipulator with several degrees of freedom capable of being programmed to carry out more than one industrial task.

8.4/1   Robot configurations

Industrial robots can perform manoeuvres within a 3-dimensional working envelope and the extent of this working envelope will be dictated by the configuration, or geometry, of the robot employed. Four **robot configurations** can be identified.

1   *Cartesian*
This configuration is illustrated in Fig. 8/6*a*. It may also be referred to as a *rectangular* configuration. The three perpendicular axes will generate a 3-dimensional rectangular-shaped working envelope.

2   *Cylindrical*
This configuration is illustrated in Fig. 8/6*b*. The geometry combines vertical and horizontal linear movement with rotary movement in a horizontal plane. As its name implies, a cylindrical working envelope will be generated by this configuration.

3   *Polar*
This configuration is illustrated in Fig. 8/6*c*. It may also be referred to as a *spherical* configuration. Rotational movement in both horizontal and vertical planes is combined with a single linear movement of the arm. The working envelope will be largely spherical but with a conical-shaped "dead zone" above the base due to the inability of the geometry to "reach" this area.

4   *Articulated*
This configuration is illustrated in Fig. 8/6*d*. It may also be referred to as an *angular* configuration. Jointed movements are provided that approximate to the joints of the human body. Terms such as waist, shoulder, elbow and wrist are used to describe the movements and indicate which joint is operative. The working envelope will be spherical.

Whilst the above configurations imply only three degrees of freedom, the full six degrees of freedom are usually accommodated by an end effector possessing the three "missing" degrees of freedom.

All types of industrial robot need to be programmed. There are predominantly three modes by which this may be accomplished:

1 *Walk-through Programming*

In this mode, the robot is moved by remote control to the various positions it must visit. This is commonly done via a key-pad or *teach pendant*. At each position, a "teach" button is pressed to retain the position in the control system memory.

2 *Lead-through Programming*

Often referred to as "lead by the nose" programming. This system involves the operator physically guiding the robot through the desired path and sequence of events. The control system stores the entire movement sequence and reproduces it faithfully. This mode is ideally suited to applications such as paint spraying where the experience and dexterity of the human operator need to be reproduced.

3 *Off-line Programming*

This mode can be used to program the movements and actions of the robot without it necessarilly being caused to traverse the desired path. It is also often used as a convenient means of editing or modifying existing sequences. It is usually accomplished via a computer using a specialised robot programming language.

8.4/2   Robot generations

Robot technology, like electronic technology, is developing in distinct phases or *generations*. At present, three generations can be identified:

First-generation Robots can be likened to devices operating under open loop control. They invariably work through a preprogrammed sequence of operations whether work is present or not. They cannot detect any change in the surrounding environment and cannot therefore modify their actions accordingly.

Second-generation Robots are robots equipped with a range of sensors, and the necessary computing power, to modify their actions due to small changes in the surrounding environment. For example, proximity (touch) sensors can detect the presence or absence of components and initiate the robot cycle accordingly. Vision sensors can differentiate between a number of components and "instruct" the robot to execute a different sequence of events depending on the component identified.

Third-generation Robots, at present only in the research stage, will be characterised by their ability to plan, make decisions, and execute tasks "intelligently". They are likely to be programmed to operate so as to maximise some defined objective.

First-generation robots form the great majority of industrial robots in current use. Second-generation robots are under continual development and used in limited numbers. Third-generation robots are still in the research stage. Development of third-generation robots will depend to a large extent on parallel developments being made in Artificial Intelligence (AI) software systems.

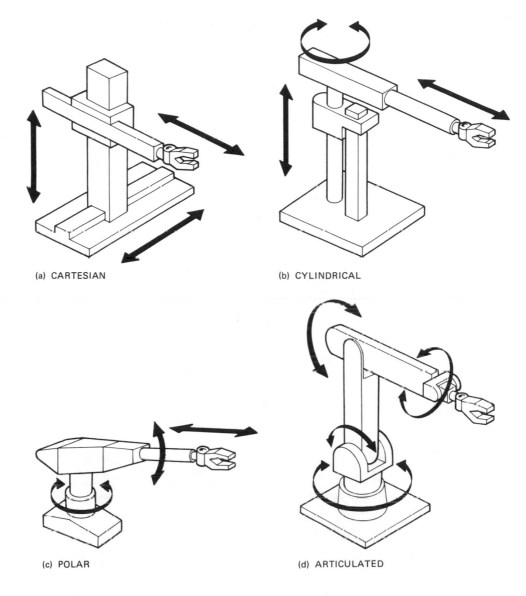

Fig. 8/6 Robot configurations

(a) CARTESIAN

(b) CYLINDRICAL

(c) POLAR

(d) ARTICULATED

8.4/3 Industrial applications of robots

The industrial applications of robots are many. Certain factors need to be considered before applying robots to a specific task. These include: reach, working envelope, lifting capacity, gripping capacity, accuracy, manoeuvrability, accessibility, repeatability, speed, form of motion (point-to-point vs. continuous path), and the provision of sensory feedback. The more common applications include:

Loading and Unloading of Manufacturing Processes
  press and m/c tools, injection moulding and die casting machines, conveyors, furnaces, etc.
Palletising and Packaging Processes
  packing and loading of components onto pallets

Welding and Flame Cutting Processes
   seam, spot and continuous run
Coating Processes
   dip and spray applications
Fettling and Cleaning Processes
   steaming, blasting, grinding, polishing, deburring
Assembly Processes
   pick and place, store and retrieve
Applicating and Dispensing Processes
   pouring, scooping, ladling—adhesive, mastic, liquid, powder
Testing and Manoeuvring Processes
   X-ray, radioactive environments

Robots are employed in industrial applications for one or more of the following reasons:

a) They can carry out boring and repetitive tasks with accuracy and consistency, thus freeing human operators from mindless tasks.

b) They can replace human operators in dangerous, hostile or obnoxious environments, thus eliminating unhealthy working conditions.

c) They can achieve flexible operation and savings in manpower by being able to operate at unsocial hours and requiring minimum services resulting in significant cost savings.

d) They are stronger and have greater reach than human operators.

e) They can achieve consistent, predictable quality and productivity for sustained periods of time.

f) They are cheaper to employ and run than human operators.

g) They offer a low-cost route into Flexible Manufacturing.

Robotic assembly of bathroom shower bases
(Courtesy: Dainichi Sykes Robotics Ltd)

Questions 8

1 Define the following abbreviations: CAD, CAM, FMS, CAPP, CAD-MAT, CIM and CAPM.
2 What is a Flexible Manufacturing System and what features distinguish it from conventional layouts of production?
3 Discuss the advantages and disadvantages of installing FMS.
4 Explain the importance of component families.
5 Discuss *two* features found in automatic work transport systems. Outline any advantages and disadvantages of employing this approach over conventional methods such as conveyor systems and fork lift trucks.
6 How are AGVs guided on the factory floor?
7 What is an ASR warehouse?
8 Describe the common elements and features of a typical CAD system.
9 What influence has ergonomics had on CAD installations?
10 How does CAM integrate with CAD?
11 What is an engineering database and who would use it?
12 Explain the need for, and the concept of, adaptive control.
13 Outline *three* ways in which tool wear can be sensed.
14 Describe how broken tools can be detected during automatic machining cycles. Discuss the possible courses of action that should be taken when a broken tool is detected.
15 Define the term "industrial robot" and outline what features differentiate them from other forms of automatic device.
16 Explain, illustrating your answer with neat sketches, *four* different robot configurations. In each case suggest a task where each robot configuration would be: *a*) suitable and *b*) unsuitable.
17 State and explain *three* means of programming industrial robots.
18 What is meant by the term "robot generations"?
19 State *six* industrial applications for which robots can be employed.
20 State *four* reasons for adopting robots for industrial tasks.

Programming examples 9

9.1 Introduction

9.1/0 Control systems

Readers attempting the following programming examples should follow the programming procedure for the CNC control system with which they are most familiar. The solution to each assignment should make reference to the particular machine type, control system and part programming format used. A nucleus of common "standard" G and M codes are presented in Chapter 7, section 7.1/4.

Additional or "non-standard" codes, specific to a particular control system, should be explained in the documentation that must accompany the solution.

9.1/1 Tooling

Select appropriate cutting tools for milling operations unless they are specified in the particular programming example. For turning operations give prior thought to compiling a tool library suitable for loading the tool turret on a CNC turning centre. Between 8 and 14 tools should prove sufficient. Certain information will need to be contained within the tool library documentation, for example type and size of tool and typical qualifying dimensions. Tools that form a typical turret set-up may include the following:

1. Right-hand Turning tool.
2. Left-hand Turning tool.
3. Finish Turning tool.
4. Face to centre—Facing tool.
5. Face from centre—Facing tool.
6. Right-hand Boring tool.
7. Left-hand Boring tool.
8. Outside-diameter Grooving tool.
9. Inside-diameter Grooving tool.
10. Face Grooving tool.
11. External V-thread—Threadcutting tool.
12. Internal V-thread—Threadcutting tool.
13. Drill.
14. Drill.

9.1/2 Documentation

All part programs should be accompanied by associated documentation. Examples of essential documentation are presented in Chapter 5, section 5.3/1. Such documentation should be considered as part of each programming example and should *not* be omitted. The self-discipline required to produce accurate and complete documentation must be developed as a prerequisite to becoming a successful part programmer.

9.2 Approaching programming

9.2/0 Planning

Prior to coding the part program for any of the programming examples, reference should be made to Chapter 5, section 5.1/2. The full planning sequence suggested in this section should be carried out. If, as a result of following this procedure, certain points are identified as missing or incomplete, assumptions must be made for convenience. This should be encouraged as an essential part of developing confidence and assuming responsibility for decisions taken by the programmer. Any assumptions made should be clearly stated within the documentation that accompanies the finished part program.

9.2/1 Programming

The appearance of a fully dimensioned, often complex component may at first sight be somewhat overwhelming. The programmer should acquire the skill of looking at such tasks as a series of clearly defined elements. For example:

Starting the program
Tool positioning (no cutting) — point to point moves
Straight line machining — linear interpolation
Circular arc machining — circular interpolation
Common operations — canned cycles
Repeated features 1 — loops, subroutines, macros
Repeated features 2 — reflection, rotation, translation
Tool changing
etc.

Looked at in this way, complex machining tasks can be visualised in a manageable way. A part program then becomes the combination of these clearly defined features.

Reference back to Chapter 7 will give the necessary support in the application of these features. Where possible always use a pre-printed part programming sheet (see Fig. 7/2).

Cutter paths must then be checked for possible collision conditions and the program verified. Refer to Chapter 7, section 7.5 for the various methods of proving part programs.

9.2/2 Example 1

Fig. 9/1 shows a component which is to be machined on a CNC milling machine with manual tool change. Prepare a program, in absolute mode, to produce: the aperture, the six tapped holes, and the two dowel holes.

Notes
The component is supplied machined to size all over.
Do not take any roughing cuts or use program cycle files.
The material is mild steel.

Fig. 9/1

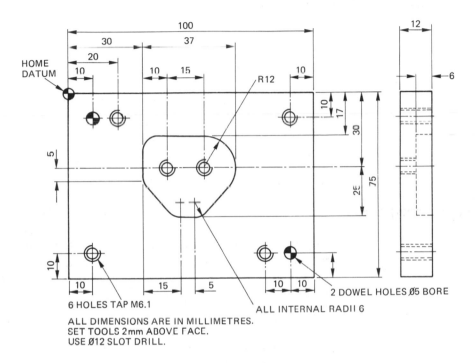

6 HOLES TAP M6.1
2 DOWEL HOLES Ø5 BORE
ALL INTERNAL RADII 6
ALL DIMENSIONS ARE IN MILLIMETRES.
SET TOOLS 2mm ABOVE FACE.
USE Ø12 SLOT DRILL.

9.2/3 Example 2

The component of Fig. 9/2 is to be machined from a premachined block (100 × 80 × 15 mm) held in a vice on a CNC milling machine with manual tool change.

a) Write a part program, using absolute units, to
    (i) Mill the annular groove using a 10 mm diameter slot drill.
    (ii) Spot drill, drill, and counterbore the four holes.
    (iii) Drill and bore the 20 mm diameter hole.
b) List tooling you have used, together with the relevant speeds and feeds.

Notes
The material is Aluminium alloy.
The Z datum is on the face.

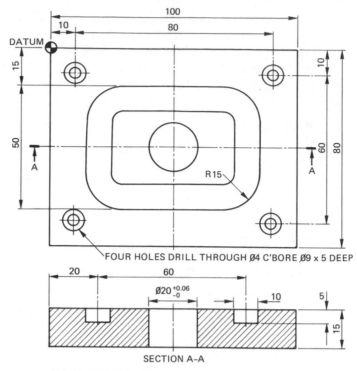

FOUR HOLES DRILL THROUGH ⌀4 C'BORE ⌀9 x 5 DEEP

SECTION A–A

ALL DIMENSIONS ARE IN MILLIMETRES

9.2/4 Example 3

Fig. 9/3 shows a die aperture to be machined from a premachined block held in a vice on a CNC milling machine with manual tool change. Write a part program to

a) Mill out the aperture using a 10 mm diameter slot drill.
b) Drill and ream the dowel holes.
c) Drill and tap the 4 tapped holes.
d) The material is mild steel.

9.2/5 Example 4

The component shown in Fig. 9/4 is to be machined from a low-carbon steel plate which is approximately 2 mm oversize on the profile and has been previously machined to produce the two locating dowel holes, and the clamping hole. It is held in a fixture which locates and clamps the component. The CNC machine tool has a manual tool change.

a) Write a part program using absolute units to
   (i) Mill the profile using a 16 mm diameter end mill.
  (ii) Mill the 10 mm slot.
 (iii) Spot and drill the four 4 mm diameter holes.
b) List the tooling used together with suitable speeds and feeds.

Notes
The Z⌀ datum is on the top face of the component.
Cutter diameter compensation may be used if desired.
Safe clearance plane is 15 mm above the surface.

Fig. 9/3

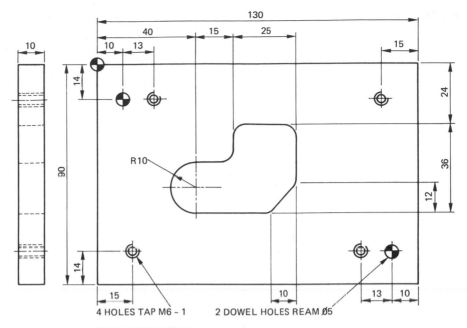

4 HOLES TAP M6 – 1   2 DOWEL HOLES REAM Ø5

PLATE THICKNESS 10mm
ALL UNDIMENSIONED RADII ARE 5mm
ALL DIMENSIONS ARE IN MILLIMETRES

Fig. 9/4

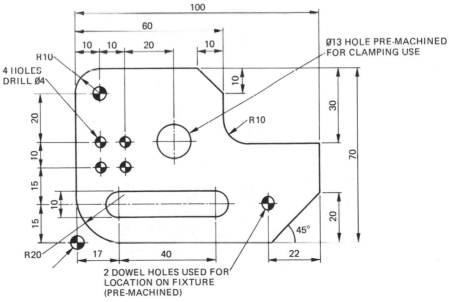

Ø13 HOLE PRE-MACHINED FOR CLAMPING USE

4 HOLES DRILL Ø4

2 DOWEL HOLES USED FOR
LOCATION ON FIXTURE
(PRE-MACHINED)

ALL DIMENSIONS ARE IN MILLIMETRES

MATERIAL 8mm THICK

9.2/6 Example 5

Fig. 9/5 shows a turned component held in soft jaws on a premachined register of 80 mm diameter, which is to be machined on a CNC turning centre. The dotted lines indicate a previous roughing-out operation and the 10 mm hole has been drilled.

a) Write a part program to finish the component to the dimensions specified.
b) List the tooling used with suitable speeds, feeds and assumed preset values.

Notes
The material is free cutting low-carbon steel.
Assume point programming (do *not* use tool nose radius compensation).
  Only program *two* passes to cut the screw thread. Ignore the acceleration and deceleration calculations which would normally be required on some systems.

Fig. 9/5

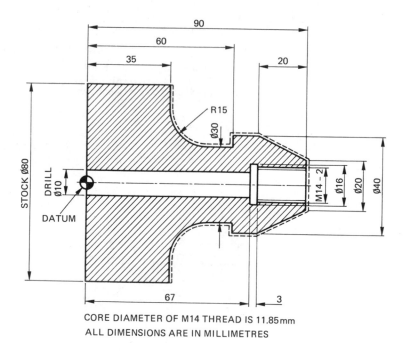

CORE DIAMETER OF M14 THREAD IS 11.85 mm
ALL DIMENSIONS ARE IN MILLIMETRES

9.2/7 Example 6

Fig. 9/6 shows a turned component held in soft jaws on a premachined register of 90 mm diameter, which is to be machined on a CNC turning centre. The dotted lines indicate a previous roughing-out operation.

a) Write a part program to finish the component to the dimensions specified.
b) List the tooling you have used, together with assumed preset values.

Notes
Assume point programming (do *not* use tool nose radius compensation).
The material is Aluminium alloy.

Fig. 9/6

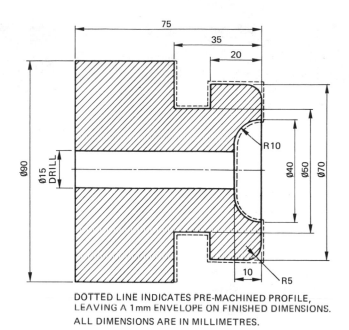

DOTTED LINE INDICATES PRE-MACHINED PROFILE,
LEAVING A 1mm ENVELOPE ON FINISHED DIMENSIONS.
ALL DIMENSIONS ARE IN MILLIMETRES.

Examples 1–6 are reproduced courtesy of the City & Guilds of London Institute

Appendix A

ISO and British Standards having direct relevance to CNC machine tools
(listed in ISO Standard number order)

Topic	ISO Standard	British Standard
ISO 7-bit coded character set for information interchange.	646:1983	4730:1980
N.C. machines: Axis motion and nomenclature.	841:1974	3635:1972* Part 1
Representation of 7-bit coded character set on punched tape.	1113:1979	3880:1972 Parts 3/4
Punched tape: Dimensions and location of feed and code holes.	1154:1975	3880:1971 Part 2
Use of longitudinal parity to detect errors.	1155:1978	4505:1981 Part 3
Character structure for start/stop and synchronous transmission.	1177:1973	4505:1981 Part 2
N.C. machines: Vocabulary.	2806:1980	6135:1981
N.C. machines: Symbols.	2972:1979	3641:1980 Part 2
N.C. machines: N.C. processor output.	3592:1978	5110:1979 Part 1
N.C. machines: Specification of interface signals: unit/machine.	4336:1981	—
N.C. machines: N.C. processor output.	4343:1978	5110:1978 Part 2
Interface specifications between numerical controls/industrial m/cs	5782:1979	—
N.C. machines: Operational commands and data format.	TR/6132:1981	—
N.C. machines: Program format and definition of address words.	6983:1982 Part 1	3635:1972* Part 1

* Standard withdrawn May 1985.

Appendix B

ISO and British Standards having direct relevance to CNC Tooling
(Listed in ISO Standard number order)

Topic	ISO Standard	British Standard
Turning tools with carbide tips: External tools.	243:1975	
Turning tools with carbide tips: Designation and marking.	504:1975	
Carbide groups and applications for machining.	513:1975	
Turning tools with carbide tips: Internal tools.	514:1975	
Specification for milling cutters: Recommended outside diameters.	523:1974	122:1980 Part 3
Indexable carbide inserts: Without fixing hole.	883:1976	4193:1980 Part 2
Indexable (throwaway) inserts for cutting tools: Designation.	1832:1977	4193:1980 Part 1
Indexable carbide inserts: Cylindrical fixing hole.	3364:1977	4193:1980 Part 3
Indexable carbide inserts: for milling cutters: Square inserts. Triangular inserts.	3365:1977/80 Part 1 Part 2	4193:1980/2 Part 4 Part 5
Turning/copy tool holders and cartridges for indexable inserts.	5608:1980	4193:1982 Part 6
Single point turning/copy tool holders for indexable inserts.	5610:1981	4193:1982 Part 7
Cartridges for indexable inserts: Dimensions.	5611:1981	4193:1982 Part 8
Boring bars for indexable inserts	6261:1983	4193:1984 Part 14
End mills with indexable inserts: Flatted parallel shank. Morse taper shank.	6262:1982 Part 1 Part 2	4193:1982 Part 9 Part 10
Face milling cutters: With indexable inserts.	6462:1983	4193:1983 Part 11
Side and face milling (slotting) cutters with indexable inserts.	6986:1983	4193:1984 Part 12
Indexable carbide inserts with partly cylindrical fixing hole.	6987:1983 Part 1	4193:1984 Part 13

Appendix C

Units and Prefixes

When units become very small or very large, multiples or sub-multiples of 10 are often used. Special prefixes are assigned to each multiple or sub-multiple, and each prefix has its own symbol. The prefixes are:

Prefix	Symbol	Multiplies by	Example
exa	E	10^{18}	
peta	P	10^{15}	
tera	T	10^{12}	
giga	G	10^{9}	$1\,GN = 1\,000\,000\,000$ Newtons
mega	M	10^{6}	$1\,MW = 1\,000\,000$ Watts
kilo	k	1000	$1\,km = 1\,000$ metres
hecto	h	100	$1\,ha = 100$ ares
deca	da	10	
—	—	—	—
deci	d	0.1	
centi	c	0.01	$1\,cm = 0.01$ metre
milli	m	0.001	$1\,ml = 0.001$ litre
micro	μ	10^{-6}	$1\,\mu s = 0.000\,001$ second
nano	n	10^{-9}	
pico	p	10^{-12}	
femto	f	10^{-15}	
atto	a	10^{-18}	

The prefix should be selected such that the number preceding it is between 0.1 and 1000. For example: 100 megawatts, NOT 100 000 kilowatts.

The following general principles should be followed when using SI units:

1 SYMBOLS for SI units are the same for both singular and plural. The letter "s" must NEVER be added to form a plural.
2 If the unit is WRITTEN out in full then the normal rules of grammar apply, e.g. 1 kilogramme, 5 kilogrammes.
3 Very large or very small numbers are often broken up, for easy reading, by grouping digits in threes to the left or right of the decimal point, e.g. 560 000.01 and 0.000 5.
4 When expressing a quantity less than unity (1), a nought should always precede the decimal point, e.g. 0.375, NOT .375.

Glossary of terms

Absolute
Absolute Coordinates: all dimensions are referenced from a fixed point.
Absolute Programming: all axis movements are specified in relation to a fixed (datum) point.

Adaptive Control Continuous monitoring and automatic adjustment during machining of various parameters, to maintain optimum cutting conditions.

Alphanumeric A character which is either a letter of the alphabet or a numerical digit.

Analog Quantity A quantity that varies continuously with time.

Analog-to-Digital Convertor (ADC) Electronic component that gives a digital number as an output, which is proportional to a varying voltage it receives as an input.

ASCII Code American Standard Code for Information Interchange. Standard coding system of 7-bit data for alphabetic, numeric, punctuation and special control characters for data communication and storage.

Automatic Pallet Changer (APC) Automatic system of changing components on CNC machining centres, coordinate measuring machines, etc. Components are loaded onto interchangeable sub-tables called pallets.

Automatic Store and Retrieve Warehouse (ASRW) Computer-controlled warehouse in which components are automatically stored in, and retrieved from, a rectangular matrix of storage locations.

Automatic Tool Changer (ATC) A mechanical arm-like device integral with the machine tool, for the purpose of automatically changing cutting tools under the control of a part program. The new tool is selected from a pre-prepared tool magazine.

Automatically Guided Vehicle (AGV) A self-contained free-ranging driverless vehicle whose route can be modified by remote or on-board computers. Used to load, unload and transport various cargos within an overall materials handling system.

Axis Direction of motion of a particular machine movement. May be either linear or rotary and is identified by a letter of the alphabet.

Backing Store Mass storage devices or media for the permanent storage of information, programs and data. Usually punched tape, magnetic tape or magnetic disc for CNC applications. Characterised by relatively long access times.

Baud Rate Bits per second. An expression of data transmission speed between two computer devices connected for serial transmission. As a rule of thumb, (baud rate/10) gives the transmission speed in characters per second.

Binary A numbering system employing two digits 0 and 1 that forms the basis of digital computer design and operation.

Binary Coded Decimal (BCD) A special form of binary code in which each decimal digit is expressed as a 4-bit binary pattern. Large or small decimal numbers can then be translated by coding each decimal digit, forming the number, in turn.

Bit Contraction of BInary digiT. A single digit in a binary number. Each bit in a binary number has a weighting dependent on its position. The rightmost bit is termed the Least Significant Bit and the leftmost bit is termed the Most Significant Bit.

Black Box A concept of visualising the operation of any system by considering just the input(s) to and the output(s) from that system. Since it is not necessary to know how the system works (just how it affects the inputs to produce the outputs), the system is known as a black box.

Block The term given to describe one line of a part program. A block consists of all the information required to carry out one machine operation. Each block is identified by a unique block or sequence number and is terminated by an "end of block" character.

Block Diagram A simplified means of representing the overall operation of a system on paper. The elements of the system are represented by symbols. The interconnection, inputs and outputs of the various elements are represented by straight lines. The direction of information flow or movement is indicated by arrowheads on the interconnection lines.

Buffer   A temporary holding store between two stages of a process that are operating at different speeds. In a CNC control unit it is an area of random access memory through which data passes when being transferred from one device to another.

Byte   A group of 8-binary bits. Each byte is capable of decoding one character of information. Computer memory capacity is measured by the number of characters (bytes) it can store. The term *kilobyte* or K represents 1024 bytes and is used as an expression of memory size.

CAD/CAM   Computer Aided Design/Computer Aided Manufacture. General terms used to describe the application of computer technology to the engineering functions of producing engineering drawings and components.

Canned or Fixed Cycle   A fixed sequence of machine operations brought into action by a single preparatory G function. Each fixed cycle command must be accompanied by additional information.

Cartesian Coordinates   A system of coordinates where features are located by dimensions at right angles to each other. Also known as *rectangular* coordinates.

Character   A single unit of information. May be a digit, letter of the alphabet, punctuation symbol, graphics symbol, or control symbol that can be represented as one byte of information.

Checksum   An error-checking device for transmitted data. Each byte of data is manipulated arithmetically to produce a coded check digit which is transmitted along with the data. The device receiving the data also decodes the check digit. Any difference between the two values indicates an error in transmission.

CL Data   Cutter Location data. Numerical information describing a tool or cutter path which is independent of any particular machine tool. Usually generated by a computer-assisted part programming system as a result of processing a geometrical shape definition. CL Data is converted into part program code for a specific machine by a post-processor or link.

Closed Loop Control   System of machine control in which various output conditions (e.g. slide position and velocity) are continuously monitored and compared with the input (command) signal. The monitoring and comparing operations are termed *feedback*.

Compensation   An automatic correction to a controlled axis that occurs at right angles to the programmed cutter path. Used to compensate for cutters of different diameter to those specified in the original part program. The amount of compensation is manually entered into the CNC control unit prior to commencing machining.

Component Family   A number of different components exhibiting broadly similar characteristics of shape and function enabling them all to be produced on a closely defined set of machines.

Computer Aided Part Programming (CAPP)   A programming system where a computer generates finished part program code from a component shape definition.

Computer Integrated Manufacture (CIM)   A total design concept of a computer-controlled manufacturing environment. An essential feature of CIM is that a single source of manufacturing data (manufacturing database) can be accessed and used by all the different functions of the organisation.

Computer Numerical Control   A system of controlling machine tools by coded, alphanumeric instructions contained within a dedicated stored-program computer control unit.

Constant Surface Speed   A control system function that automatically adjusts the machine spindle speed as component size increases or decreases. Normally employed on CNC turning centres. CSS is activated within a part program by a preparatory (G) code.

Control Engineering   An engineering discipline dealing with the design and operation of automatically controlled systems.

Conversational Programming   CNC control system software that allows a machine tool to be programmed by the operator entering information in response to a number of questions provided by the machine operating system.

Crash Zone   An operator-defined dimensional boundary that can be used to restrict slide or table movement. If a programmed tool path attempts to enter this boundary, the control system causes the machine to cease operation and generates an alarm signal to alert the operator. There may be different levels of crash zone as provided for by the control system, e.g safe, warning and fault.

Damping   Resistance to motion introduced to reduce undesirable oscillations and counter the effects of excessive overshoot or undershoot.

Datum   Fixed reference point about which movements or measurements are made.

Dedicated Computer   A computer system dedicated to carrying out a fixed, clearly defined task.

Detailed Format Classification   A classification that details the order and form of commands, dimensions and other information that can be accepted by a particular CNC control system.

Digital Quantity   A quantity that can be represented by one of two mutually exclusive states. Since it does not matter what actual values represent the two states they are referred to as 1 and 0.

Digital-to-Analog Convertor (DAC)   Electronic component that gives, as an output, a varying voltage which is proportional to a digital number it receives as an input.

Dimensional Tolerance   The amount of permitted error allowed (either above and/or below) from a stated dimension.

Direct Numerical Control   The control of one or a number of CNC machine tools directly by a host computer. The downloading of part programs from a host computer directly into the memory of a CNC machine tool.

Editing   Facility whereby the machine operator can modify a stored program and/or information registers, in the memory of the CNC control unit, without altering the original program medium from which the stored program was read.

EIA Coding System   U.S. Electronic Industries Association standard coding system for the representation of characters on 8-track punched tape. Track 5 is used for parity checking.

EPROM   Erasable Programmable Read Only Memory. A programmable memory chip whose contents remain intact even when the power source is removed (non-volatile). The memory chips can be erased by exposing them to ultra-violet light for a short period of time, and then re-programmed.

Feedback   A feature of closed loop control systems whereby an electrical signal, proportional to a monitored quantity (axis position or velocity), is compared with the command signal. The result of the comparison is an indication of the error between the commanded and the actual condition.

Firmware   A general term used to describe programs (software) stored on a memory chip.

First-Off   Term given to the first machined component inspected to verify that the part program used to machine the component is correct.

Fixed Zero   A fixed point on a numerically controlled machine tool about which all machine movements are referenced.

Flexible Manufacturing System   A means of organising for production using high levels of automated equipment and computer control to achieve largely unmanned manufacture.

Format   See *Part Program Format.*

Full Floating Zero   Feature of a CNC control system that allows the machine zero datum to be positioned anywhere within the programmable area of the machine, for convenience.

Gain   Design function of a control system which is a measure of the amplification required between input and output signals to balance the behaviour of the system.

Gauge Height or Gauge Plane   The height or plane above the surface of the workpiece at which rapid travel of the spindle in the Z-axis changes to feed.

Geometrical Tolerance   The amount of permitted error in shape or form, from true geometry.

Grating   A linear or circular component comprising a large number of equally spaced lines ruled either on glass or bright metal. When ruled on glass, the optical grating produced consists of alternate transparent and opaque regions. When ruled on bright metal, alternate light and dark bands are sensed by reflective techniques.

Grid Plate   A sub-plate comprising accurately machined features and tapped securing holes, used for the quick and accurate location of components to be machined.

Handshaking   A system used between two interconnected devices that operate at different speeds, to ensure that all data transmitted by the transmitting device is received correctly by the receiving device. Control signals signalling "ready to receive" and/or "ready to transmit" must be present to initiate data transfer.

Hardmetal Inserts   Sintered carbide cutting tool inserts of various shapes and sizes that may be clamped into a variety of holders and cartridges to form cutting tools. More than one cutting edge is normally provided on each insert and they are discarded (not reground) when exhausted. May also be referred to as *indexable inserts* or *throwaway tips.*

Hardware   A general term used to describe all physical components of a computer system.

Hunting   Feature of a closed loop control system whereby a moving axis slide oscillates about its commanded destination position. Caused by a poorly designed closed loop control system trying to compensate for the effects of excessive overshoot and/or undershoot when negotiating a target position.

Incremental
Incremental Coordinates: all dimensions are referenced from a previously dimensioned point rather than a fixed reference point.
Incremental Programming: all axis movements are specified in relation to the last point visited rather than from a fixed (absolute) point.

Inductosyn   A position-measuring transducer working on a similar principle to the synchro resolver. Often used for fine measurements.

Instability   See *Hunting.*

Interface   A collection of electronics designed to make output signals from one device compatible with the input signal requirements of another device.

Interpolation   The joining up of programmed points to generate a smooth path by computation. If the segments joining the points are straight lines, the process is called *linear interpolation*. If the segments joining the points are arcs of circles, the process is called *circular interpolation*. If the segments joining the points are arcs of a parabola, the process is called *parabolic interpolation*.

ISO Code System   International Standards Organisation standard coding system for the representation of characters on 8-track punched tape. Track 8 is used for parity checking.

Jogging   Method of manually controlling machine axis movement. The depression of a jog button will move the selected axis by a fixed amount of movement. The amount of movement per depression, the direction of movement and the axis are all selectable by the operator. May also be referred to as *inching*.

Lead Time   The amount of time that elapses between the receipt of an order and subsequent delivery of that order.

Loop   A repetition function of a computer or part program language. Allows a defined sequence of steps to be repeated a desired number of times. A means of causing the flow of a program to be transferred to a previous point in the program.

Machine Control Unit (MCU)   Term given to describe the physical components of a CNC control system. Includes the control console, VDU, and input keyboard.

Machining Cell   An organisation of a small number of machine tools arranged to manufacture a defined set of components.

Machining Centre   A CNC machine tool resembling a milling machine which is capable of carrying out a large number of machining operations in at least three axes under simultaneous contouring control. May include integral automatic tool and workpiece-changing facilities and swarf disposal systems.

Macro   A facility of a programming language that allows a specified number of commands, in the form of a sub-program, to be executed by a single program call. May include the facility to pass parameters to the sub-program.

Manual Data Input (MDI)   The manual keying-in of part programs to the memory of a CNC control unit via the console keyboard.

Manual Override   Controls on a CNC machine tool that allow the operator to override programmed feed and speed values without permanently altering programmed values held in memory. Override values of up to 120% of programmed values are generally provided.

Manual Part Programming   The manual coding of part programs using G and M codes.

Maximum Material Condition (MMC)   The condition of a component at its limit of size such as to maximise the material present within that component. For example, a round shaft will be at its MMC when its diameter is largest or at its upper limit of size. A hole will be at its MMC when its diameter is at its smallest or at its lower limit of size.

Memory   Any device capable of storing information for retrieval at a later time. See also *RAM*, *ROM* and *EPROM*.

Microelectronics   The miniaturisation of electronic circuits and components.

Microprocessor   A programmable integrated circuit (chip) that forms the heart of many general-purpose and dedicated computer control systems. Must generally be provided with external means of input and output signals and external memory to form a useful computer system.

Mirror Imaging   A part program function that allows opposite-hand features to be produced by a single programmed command. Works by reversing the sign of programmed dimensions in one or two axes. Also used to repeat, in other quadrants, features programmed in a single quadrant. Also termed *reflection*.

Miscellaneous Function   A two-digit command preceded by the letter M. Also called M-functions or Auxiliary functions. Coded commands that represent a particular switching function of the machine tool, for example spindle on, spindle off, coolant on, coolant off, etc.

Modal Function   A programmed function that remains in effect until cancelled or superseded by a function of the same type.

Moiré Fringe   A repeating interference pattern of light and dark alternating bands produced as a result of two optical gratings moving relative to each other. There is a mathematical relationship between the number of lines on the gratings and the movement of the fringes for a given movement of the gratings.

Negative Feedback   A control system concept. The comparing of a monitored output signal with a commanded input signal by subtraction. A result of zero indicates that the actual value coincides exactly with the commanded (target) value. A non-zero result indicates that the target value has not yet been reached or has been exceeded and corrective action is required.

Numerical Control (NC)   Automatic control of machine tool actions by a pre-determined program of coded instructions and commands.

Offset   An automatic correction that occurs parallel to a controlled axis. Usually operative in the Z-axis to compensate for differences in the actual lengths of cutters and the lengths of cutters originally programmed. Offset values are manually entered into the CNC control unit prior to commencing machining.

Open Loop Control   A control system in which there is no feedback. Position and velocity are determined by inbuilt features of the driving mechanism.

Operating Program   The master control program that directs the action of a CNC control system. Held in non-volatile, read only memory. It is provided by the control system designers and cannot be modified by the end user.

Overhead   An expense item that does not contribute directly to the manufacture of a component or service. Such costs have to be borne by sales of the component or service.

Parameter   A numerical value that once defined remains constant, but may subsequently be re-defined to a different value for use under the same circumstances.

Parametric Programming   A system of programming where dimensional values are replaced by letters or symbols. Actual dimensions are supplied when the program is executed. This enables the same program to be used to produce components of the same form but of different dimensions.

Parity Check   An error-checking device for binary transmitted data by automatically counting the number of bits set to binary 1. Applied to punched tape by counting the number of holes punched across the tape. The tape-punching device ensures that the number of holes punched is *always* an odd or an even number, depending on the system employed. If the device reading the tape detects an incorrect number of holes, it is assumed that a transmission error (damaged tape?) has occurred.

Part Program   A complete set of coded instructions for the complete machining of a component or part. Must be processed by the machine control unit.

Part Program Format   The order and form in which part programs must be written to be accepted by a particular control system. In general, formats may be either fixed or variable block. In fixed block format, each command word must be specified in every block. In variable block format, only those commands that change need be specified.

Peripheral   General term used to describe a physical piece of equipment used in conjunction with a computer system, for supporting functions. Common peripherals include tape readers and printers.

Photocell   An electronic component (transducer) that gives an output voltage dependent on the amount of light falling on it. Some photocells give a constant voltage once a particular light threshold has been reached, whilst others give an analog voltage proportional to the light intensity falling on the photocell.

Planetry Roller Leadscrew   A leadscrew system in which sliding friction is replaced by rolling friction. Threaded rollers revolve around a threaded leadscrew in a manner similar to the way planets revolve around a sun.

Polar coordinates   A system of coordinates in which points are described by a length from a pre-defined datum point, and an angle from a specified plane.

Positional Measuring System   Measuring system designed to monitor the position or movement of a controlled axis, to produce feedback in a closed loop control system.

Post-Processor   A computer program (software) which converts dimensional (workpiece) data, output by a computer aided part programming system, into a form suitable for the control system of a particular CNC machine tool. May also be called a *link*.

Preparatory Function   Also called a G-function. A 2-digit number preceded by the letter G. Codes that prepare the control system for a particular mode of operation. Examples include absolute/incremental coordinates, inch/metric dimensions, rapid/feed axis movement, linear/circular interpolation, etc.

Pre-set Tooling   A system of setting cutting tools away from the machine tool, on special equipment that emulates the machine datum conditions. Allows tool changes and replacement to be accomplished quickly to reduce machine downtime due to tool setting.

Printed Circuit Board   Electronic circuits produced by discrete components on self-contained boards or cards. May allow the simple replacement of damaged or faulty circuitry and a simple means of updating or enhancing control system facilities.

Program   A sequence of instructions conforming to certain rules of spelling and syntax that can be executed by a computer or microprocessor system, to perform a particular task.

Program Proving   Technique(s) of verifying the safe and correct operation of part programs running on CNC machine tools.

Punched Tape   A storage medium for the permanent storage, and loading, of part programs. Punched tape for CNC applications is 25 mm (1 inch) wide and holds 10 characters per 25 mm length. Common tape materials include paper, plastic and polyester laminates.

Qualified Tooling   Cutting tools on which the position of the cutting edge is guaranteed (within ±0.08 mm) relative to various points on the tool holder.

Random Access Memory (RAM)   Computer memory that can be written to and read from (by the computer control system) with equal ease and speed. The contents of RAM are usually lost when the power source is removed (the contents are said to be volatile). Memory contents can be retained by supplying power by battery backup when the main power source is removed. Supplied as memory chips in units of 1 K capacity. Used for holding user-supplied programs.

Recirculating Ball Leadscrew   A leadscrew system in which sliding friction is replaced by rolling friction. Hardened ball bearings re-circulate around a ground form leadscrew.

Register   A random access memory location which is internal to a microprocessor. The name given to a single random access memory location used for a specific purpose.

Read Only Memory (ROM)   Computer memory that can only be read by the computer control system. Writing to ROM has no effect. The contents of ROM are permanent and remain intact even when the power source is removed (the contents are said to be non-volatile). The contents of ROM are determined by the application and once programmed cannot be modified or re-programmed. Used for holding main control system software.

RS 232 Interface   A collection of electronics that arranges the correct protocol for serial data transmission according to EIA standard RS 232(C). Such interfaces may be designed for one-way or two-way communication.

Semi-Qualified Tooling   Cutting tools on which the position of the cutting edge can be adjusted to known dimensions relative to various points on the toolholder.

Servomechanism   An automatic control system in which position is the controlled quantity. Incorporates feedback and power amplification to enable an output quantity to follow an input (command) signal without error. Used in CNC machine tools for axis positioning.

Sequence Number   Positive integer number preceded by the letter N that identifies the separate blocks, and the relative positions of those blocks, within a part program.

Shaft Encoder   A transducer in which the (analog) angular position of a rotating shaft is converted into (digital) coded form using a digitally coded disc.

Single Step   A mode of operation of a CNC machine tool running a part program. Individual blocks of the part program are executed one at a time upon the depression of a key on the console keyboard. After a block has been executed, the machine stops and awaits further commands.

Software   General term used to describe computer programs and the media on which they are stored.

Steady State   The condition of a control system when equilibrium of movement has been reached.

Stepper Motor   A digital electric motor used in open loop CNC control systems for axis movement. The rotor (connected to the axis leadscrew) moves in small, fixed angular steps (in either direction) on

receipt of digital pulses from the control system. Rotational speed (hence feed rate) depends on the frequency of the applied pulses.

Subroutine or Sub-program   A separately defined part of a computer or part program that can be called to execute from various points in the main program. Used to simplify and shorten programs by defining commonly used sequences once only, and calling them as required.

Synchro-resolver   An electro-magnetic position transducer whose output voltage depends on the angular position of its rotor.

Teletype   An electro-mechanical typewriter device that can be used manually by means of an integral keyboard, or remotely by a computer transmitting data or an integral punched tape reader. Many teletypes have integral tape punch units that can simultaneously produce punched tapes.

Tooling System   An integrated system of cutting tool mounting that allows any cutting tool to be mounted on any one of a range of machine tools quickly and easily. Allows the rationalisation and simplification of tool stocks.

Track   A path along which coded information may be stored on backing store medium. On punched or magnetic tape, a track is a path running the length of the tape and parallel with its edge. On a magnetic disc, a track is a circular path concentric with the driving hub. Sometimes called a *channel*.

Transducer   A device that converts energy in one form into energy in another form, in such a way that the output is a known function of the input.

Turning Centre   A CNC machine tool resembling a lathe that is capable of carrying out a variety of turning, boring and drilling operations, at a single set-up and in a number of axes simultaneously. Most turning centres have programmable tool turrets in which a number of tools can be brought into action under program control, and integral swarf disposal systems.

Value Analysis   An exercise carried out on existing products to determine whether, by re-design, they can be manufactured more easily or cheaply.

Word   A group of characters representing a unit of information within a part program. An example might be an X-dimension consisting of 8-characters or a preparatory function consisting of 3-characters.

Word Address   A part programming format in which each word is preceded by a letter to identify its function.

Work in Progress (WIP)   Stocks of part-finished components awaiting further processing. High work in progress means large amounts of working capital tied up in part-finished stocks.

Zero Shift   A facility on a numerically controlled machine tool whereby the machine zero can be shifted to any point within the programmable area of the machine.

Suggestions for assignments

CHAPTER 1

1 Obtain relevant engineering journals. Research and identify manufacturers of the following CNC equipment and obtain information leaflets for typical machine types.

a) Machine tools.
b) Coordinate measuring/inspection machines.
c) Punch presses.
d) Assembly machines.

Compare the specification of each machine type and list those factors that appear to be common and those that appear significantly different. Comment on your findings.

2 In section 1.31 there is a list of *eight* suggestions as to where CNC may be profitably employed. For each of the suggestions:

a) Give an example of a component or situation where the suggestion would apply.
b) Explain why CNC could be better employed in preference to conventional machining.

State *three* instances where it would be better to employ conventional machining techniques in preference to CNC machining.

3 Examine a place of work with which you are familiar which does not at present use CNC machine tools, but you feel would derive considerable benefit by doing so.

a) Write a report to justify why the installation of a CNC machine (or machines) should be installed. Be careful to present an objective argument which acknowledges any difficulties that might be encountered.
b) Suggest a "plan of action" that might be followed to bring about such an installation.
c) State your estimate of how long it would take to have your CNC installation in full production.

4 Research a CNC installation with which you are familiar. Compile a report which outlines the following:

a) Details of *new* support functions which have had to be set up as a consequence of adopting CNC.
b) How (if at all) CNC affects other business functions within the organisation.
c) The benefits (if any) that have been realised by adopting CNC.
d) The disadvantages (if any) that have been realised by adopting CNC.

e) Any management information that is being obtained directly from the CNC process.

f) Any personal observations and suggestions that might improve the benefits to be derived from the CNC operation.

CHAPTER 2

1 Conduct a detailed examination of a particular CNC machine tool and carry out the following:

a) Sketch the configuration of the machine tool indicating the position of the spindle and axis motors.

b) Comment on the structural design of the machine.

c) Make brief notes on the motion transmission and slideway elements used on the machine.

d) Identify the axis movements on a simple sketch of the machine tool.

e) Comment on any facilities provided for swarf control and safety.

2 Examine the control console of a CNC control unit on a particular machine tool and carry out the following:

a) State the particular machine tool/control system combination.

b) Sketch, in good size and proportion, the control console.

c) Label each control and display indicator on the console and comment on the layout and design.

d) Write a brief description of the function and operation of each control element and display indicator.

3 Obtain manufacturer's literature for a CNC machine tool and explain how the following points are specified:

a) The size/capacity of the machine.

b) Speed and power characteristics of the machine.

c) Tooling capacity and mounting details.

d) Work capacity and mounting details.

Comment on any other machine-specific information you would consider important in making a choice of a CNC machine tool before purchase.

4 Make a critical examination of a particular CNC machine tool and control system and identify any features that contribute to aspects of safety. Your examination should encompass:

a) Physical design features of the machine tool itself.

b) Layout and ergonomic aspects.

c) Guarding, interlocks, fail safe and emergency stop devices.

d) Operational features such as warning and alarm indicators.

e) Programming features such as safety crash zones, etc.

CHAPTER 3

1 Research the specification for a particular CNC machine tool and tabulate the following:

a) The number of axes and type of axis control.

b) The resolution, accuracy and repeatability of the control system.

c) The RAM capacity of the control unit.

d) The spindle drive system employed and its capacity.

e) Spindle speeds available and how they are specified.

f) The axis drive system employed and its capacity.

g) Feed and rapid traverse rates and how they are specified.

h) The types of positional-measuring transducers employed.

i) The types of velocity-measuring transducers employed.

j) The presence and scope of any inbuilt diagnostics capability.

2 For a typical CNC machine tool draw a block diagram of the control system indicating how positional and velocity control is effected. Indicate on the same diagram whether the control signals to and from particular blocks are digital or analog, and the points in the system where conversion between signal types (the presence of an interface) would be required. State any assumption you have made in the choice of transducer and motor types.

3 Identify *two* components that may be produced by CNC machine tools. With reference to the components you have chosen, discuss the factors that would influence the type of control system/drive system combination required to machine the components. Conclude your report by suggesting suitable system combinations which would be capable of producing the components.

4 Compile a tabular comparison chart for comparing various positional-measuring transducers. You may choose which attributes are to be compared. Some examples are: principle of operation, digital or analog, linear or rotary, how direction of motion is detected, typical/relative accuracy, relative cost, advantages, limitations, etc. Further suggestions may be obtained from manufacturers' literature.

CHAPTER 4

1 For a typical part program carry out the following exercises and perform the calculations that follow:

a) Count the number of blocks in the part program.

b) Count the total number of characters in the part program.

i) Calculate the amount of computer memory required to store the part program. How many programs of this length could be stored in a CNC control unit having 64K of random access memory?

ii) Calculate the length of punched tape required to store the part program. Why would an actual punched tape be significantly longer than this calculated length?

iii) Calculate the amount of time it would take to transmit the part program, via DNC, using an RS 232 interface operating at a transmission speed of 3∅∅ baud. What special precautions need to be taken when transmitting data via an RS 232 interface?

iv) Calculate an average characters-per-block or characters-per-ten-blocks figure that may be used as a "rule of thumb" calculation for determining the number of characters in a part program. How could such a "rule of thumb" be useful?

2 Design a conversion chart that can be used to look up and convert ISO tape codes to EIA tape codes and vice versa. The chart should allow conversions to be made on the basis of character description, character bit pattern and decimal equivalent.

3 Produce a list of all the information required to produce turned components consisting of plain diameters and screw threads only. From this list compile a logical sequence of questions that could be used to form the basis of a question-and-answer dialogue that might be used in a conversational programming system.

The user may only respond to the questions in one of three ways as follows:
i) Choosing a single option (identified by a letter or a number from a menu of options).
ii) Answering YES or NO in response to a question.
iii) Supplying numerical information by typing digits at the console keyboard.
State any assumptions you make in designing your system.

4 Examine a CNC machine tool and its associated manuals to identify and list the following:

a) The type of CNC control unit.

b) The memory capacity of the control unit available for storing part programs.

c) The means of entering part programs into the memory of the control unit.

d) The type of backing store employed.

e) If an RS 232 interface is fitted, the data transmission settings.

CHAPTER 5

1 Select a component with which you are familiar and which is being produced by conventional machining techniques. Assume that the same component now has to be produced, in small quantities, by CNC machining techniques. Critically examine the component and suggest any design changes that need to be made for CNC techniques, or which would allow it to be produced more quickly, cheaply or efficiently by CNC. Explain the reasons for the suggestions that you make together with any additional advantages or disadvantages of incorporating the changes.

2 For *one* each of the MILLING and TURNING programming examples provided in Chapter 9, plan a toolpath route sufficient to finish-machine the component. Identify all points at which the direction of the tool changes. Tabulate these points, in sequence, together with the coordinate position involved in both ABSOLUTE and INCREMENTAL coordinates.

3 List the factors that you consider necessary to fully identify and use:

a) Tooling.

b) Workholding and datums.

c) Machining operations.

From your list of factors and giving high importance to simplicity, clarity and unambiguity, design standard forms or sheets for the above, to be used within a CNC installation.

You may quote examples of both good and bad points from existing documentation with which you are familiar, to justify your designs. State any assumptions you make in designing your documents.

4 Obtain a working drawing of a component that includes reference to geometrical tolerances. For each of the tolerance symbols used explain the following:

a) The meaning of the tolerance symbol.

b) The purpose of specifying that tolerance.

c) Likely consequences of not having specified the tolerance.

State whether primary, secondary and/or tertiary datums have been referenced.

CHAPTER 6

1 For each of the programming examples on TURNING given in Chapter 9, determine the types of inserted tool required for the finish-machining of each component. For each tool determine:

a) A suitable toolholder type.

b) A suitable insert type.

c) A suitable insert size.

d) A suitable insert grade.

e) A suitable insert clamping system.

For each tool/insert combination provide an ISO classification for the tool-holder and the insert.

2 For one of the programming examples on MILLING given in Chapter 9, design a workholding fixture for producing a number of identical components. Your design should incorporate sound principles of location, clamping and ease of use.

3 For a particular CNC machine tool with which you are familiar, obtain the operator's manual. Using the manual and your own experience produce a set of instructions, in the form of a checklist, for each of the following operations:

a) Setting axis datums.

b) Entering tool offset values.

c) Searching and editing part programs.

d) Manually operating the machine tool.

e) Changing cutting tools.

4 *a*) Produce a report justifying the need for the setting-up of a standardised system of tooling for a CNC installation comprising a number of different CNC machine tools.

b) Provide suggestions of how such a scheme could be implemented. Your suggestions should include reference to tooling requirements, tooling identification, pre-setting, transport to and from the machine tool, and control of tool movements.

CHAPTER 7

1 Make a comparison of (at least *three*) different CNC control systems and compile a comparison table that includes the following information:

a) A comparison of G-functions listed in function number order.

b) A comparison of M-functions listed in function number order.

c) How both speed and feed values are specified.

For each control system state the type of part program format employed and specify a detailed format classification. Comment on the results obtained.

2 For a CNC control system with which you are familiar compile a list of canned cycles provided by that system. For each of the canned cycles.

a) Explain, in detail, the action of the cycle.

b) Specify what additional information must accompany the cycle G-code.

Suggest any additional functions that you feel would improve the programming facilities if they could be implemented.

3 For a CNC control system with which you are familiar, identify the types of repetitive programming facilities provided by that system. These may include loops, subroutines and/or macros. For each facility, explain in detail:

a) The purpose of the facility.

b) How it is specified within a part program.

c) A typical use for the facility.

d) Any way in which the facility could be enhanced by a possible control system update.

4 For *one* each of the MILLING and TURNING programming examples provided in Chapter 9, make an estimate of the time required to perform the following tasks:

a) Plan/decide on/design workholding.

b) Plan the operation sequence.

c) Decide and fully specify tooling.

d) Calculate and tabulate coordinate dimensions.

e) Design and code the part program.

f) Fully compile a complete set of documentation.

Refer to Chapter 5, section 5.12, and then carry out the above tasks (completely and honestly!) and time how long it actually takes to complete them.

i) Compare and comment on your estimated times and the actual times, suggesting reasons for any discrepancies.

ii) Give your estimation of the complexity of the component and whether the actual times represent a good average on which costing could be based.

iii) On the basis of your results put a case either for or against the use of a computer-assisted part programming system. Your case should address wider issues such as financial, organisational issues and issues of manpower and training.

Index